YOU'RE LOOKING VERY WELL

by the same author

THE UNNATURAL NATURE OF SCIENCE

MALIGNANT SADNESS:
The Anatomy of Depression

SIX IMPOSSIBLE THINGS BEFORE BREAKFAST:
The Evolutionary Origins of Belief

HOW WE LIVE AND WHY WE DIE:
The Secret Lives of Cells

You're Looking Very Well

the surprising nature of getting old

LEWIS WOLPERT

First published in 2011
by Faber and Faber Ltd
Bloomsbury House
74–77 Great Russell Street
London WC1B 3DA

Typeset by Faber and Faber Ltd
Printed in England by CPI Mackays, Chatham

A CIP record for this book
is available from the British Library

ISBN 978-0-571-25064-6

2 4 6 8 10 9 7 5 3 1

Contents

Acknowledgements

As always I have had a lot of much-needed help. I am greatly indebted to Alison Hawkes for her editing and comments throughout my writing of the book. I am also indebted to Julian Loose, my editor, and Kate Murray-Browne at Faber for their comments and encouragement. My thanks also go to my agent, Anne Engel, and to all of those who gave up their time to talk to me.

1
Surprising

'Old age is the most unexpected of all the things that can happen to a man' – Leon Trotsky

When we are young we do not think about being old; it is simply not part of our agenda. So when we age we are not prepared for it and it can be quite a surprise. It has come as a shock to me. How can a 17-year-old, like me, suddenly be 81? The only obvious features, if I do not look in the mirror, are that I now walk so slowly that most people dash past me on the pavement, and I have retired from my university job. I find it hard to come to terms with both. But I also forget names, words and faces. I recently forgot several things I had meant to do, and was worried that this was age-related and might indicate the start of dementia. I intended to ask my psychiatrist at my next appointment related to depression. I went to see him and then laughed when I remembered, as I cycled home, that I had forgotten to ask him.

I never thought much about ageing, but Shakespeare's memorable description in *As You Like It* doesn't offer a particularly positive view:

> All the world's a stage
> And all the men and women merely players;
> They have their exits and their entrances;
> And one man in his time plays many parts,

His acts being seven ages. At first the infant,
Mewling and puking in the nurse's arms;
Then the whining schoolboy, with his satchel
And shining morning face, creeping like a snail
Unwillingly to school . . .
The sixth age shifts
Into the lean and slipper'd pantaloon,
With spectacles on nose and pouch on side,
His youthful hose, well sav'd, a world too wide
For his shrunk shank; and his big manly voice,
Turning again toward childish treble, pipes
And whistles in his sound. Last scene of all,
That ends this strange eventful history,
Is second childishness and mere oblivion,
Sans teeth, sans eyes, sans taste, sans every thing.

I decided I wanted to understand ageing, not least because it seems so different a topic from the one I have spent most of my life studying, the development of the embryo. Is ageing part of our developmental programme? We are essentially a society of cells, so what is the cellular basis and what is the cause of both physical and mental decline in old age? Can these be prevented? Is immortality in the future a real possibility? As the proportion of elderly people in society rises, there are more general social and economic problems related to their treatment and care. I needed to know how the old can live well, and whether, when the time comes, we can choose when to die. Also, does understanding ageing help with one's own ageing?

Yet old age is by no means easy to define. It is a biological phenomenon characterised by certain physical changes that take place with time and also by their psychological consequences. It is the changes in individual appearance that prompt your friends to remark, 'You're looking very well.' Indeed, one can now think of there being four ages, rather than Shakespeare's seven, in our lives: childhood, active adulthood, maturity and finally 'You're looking very well.' As an 81-year-old, I hear it all the time, and use it again and again when I meet my ageing friends. There is a quite wide belief that old age starts at around 65, which has been a common age for compulsory retirement. But now that there are so many aged over 75 and even 85, we need to look again at when old age really begins. Of course ageing really begins, as we shall see, when we are quite young – there are few 40-year-old football or tennis champions.

Getting old is often more apparent to others than to the actual person who is ageing, and many who are old persist in the belief that they are still young. It is important to realise that everyone ages differently, depending on their circumstances. The big question is how one deals with old age. The author Doris Lessing, for example, finds it acutely irritating. Does one try to find new activities? Should one reflect on one's life to decide if it has been worthwhile? And how does one deal with both bodily and mental decline?

For many, ageing is frightening. Only about one in ten of those aged 75 to 79 remains free of physical illnesses such as those of the heart, eye and bones. Those with wealth and a good education do best, as do those who have a positive attitude to ageing. But though all our bodily functions deteriorate with age – for example, muscles lose their strength and the immune system weakens – evidence from those who play sports shows that even when old it is possible to continue to play quite well. There are also many false ideas about the decline of sexual activity in the elderly: in fact there is little evidence for a significant age-related decline.

There is a significant chance of developing a mental disability after the age of 50, dementia, particularly Alzheimer's disease, being the most common. This costs the UK an estimated £17 billion each year, and is a great problem for carers. Other common age-related mental diseases include Parkinson's disease and depression. But while our mental abilities undoubtedly decline with age – we become forgetful and slower – our acquired knowledge, fortunately, seems to remain intact. Many of the old hold high positions and remain creative, with their intellectual abilities in good shape.

For some, old age has very positive aspects. It can even be a time of joy. Today the title of the Beatles song would be 'When I'm Eighty-four'. At the age of 106, a neighbour of mine is very happy and still active with her piano playing. Things that are dear may become dearer, such as ideals, friendships and family. There may be time to do things which we could not do when we were young – and there is the possibility of being adventurous either physically or mentally, even if immobility is creeping on. For me it is very encouraging to see that even when old, scientists can still be very active. A fine example is Professor Dennis Mitchison, a friend of mine who is distinguished for his work on tuberculosis. Mitchison is 90 and is an Emeritus Professor at St George's Hospital medical school. I asked him about his views on ageing:

> I did not begin to notice the effects of age till I was about 85. I got a bit slower and my organs worked less well. I enjoy my current age since the research is very exciting and even if a bit

> slower I just get on with it. I collaborate and get grants mainly from drug companies. We have seven scientific papers in preparation, one of which has just been rejected but at my age I get less upset when things go wrong. We have a project that will only be finished in about ten years – not sure I will still be here. I do think that euthanasia raises difficult problems but I do think we have the right to choose when to die.

We live longer today than at any time in history. Over a sixth of people living in the UK are expected to celebrate their 100th birthday. Mortality has been considerably postponed, as a result not of revolutionary advances in slowing the process of ageing, but of progress in improving health. Ageing itself has undergone a dramatic change over recent years. In the industrialised world in the twentieth century there was an unexpected and unprecedented growth in the older population – some 30 years were added to life, an increase greater than in the previous 5,000 years. This was due to improved healthcare, food and sanitation. There are now more people aged over 65 than under 16 in the UK. And the number of people aged 85 years or more doubled between 1983 and 2008.

In the UK, there are currently around 10 million people who are over 65 and 1.3 million over the age of 85, of whom 422,000 are men and 914,000 women. Women have outlived men throughout history. We humans generally live longer than our ape-like relatives, and have an extended period of juvenile dependence – this may be related to getting food, which is difficult for the young. The elderly are, in this sense, repositories of knowledge. But while there have been many individual exceptions – such as St Augustine, who lived to 75, and Michelangelo, who died at 88 – for most of human history the average lifespan was short. In London in 1800 you could expect to live to just 30, in 1900 to 42, in 1950 to 61, and now to about 80, women for a few more years than men.

The recent increase in the number of the old and very old has major implications for how they live and are cared for. Ageing can thus have major economic impacts, particularly if the number of the aged becomes greater than the number of the young needed to support them. It is thus essential that we understand both the biological basis of ageing and how the old are treated.

Why do we age? The Ancients mainly thought that it involved the loss of some key factor in the body. The scientific study of ageing began with Francis Bacon in the seventeenth century, but it was only after cells were recognised as determining how bodies functioned that scientific research into ageing could properly progress. Even doctors eventually became interested in ageing, and geriatrics was established. Peter Medawar proposed an evolutionary theory, and there was the surprising discovery that cells aged in culture. Evolution and sex play key roles in understanding why we age: evolution is concerned only with reproduction, so does not care if we age after having successfully reproduced. Ageing is not programmed in our genes like normal growth; on the contrary, there are genes which try to prevent it. Another surprise is where our great progress in understanding has come from: it has been the investigations of ageing in a simple nematode worm and flies, as well as in mice, that have been so productive.

These studies have identified some of the molecular mechanisms responsible for ageing, and even raise the possibilities of extending life still further, but whether this is desirable unless age-related disabilities can be avoided is a key question. How long could we live? How long *should* we live? There are many myths about humans living to a very old age, but there is no evidence for any of those claims beyond 115 years for men and 122 for women. Genes can account for about one third of lifespan. In spite of much publicity and advertising, there is no known method of extending lifespan other than through exercise, not being overweight, and being healthy and positive. In model organisms like worms and flies, it is possible to increase their age fivefold, but at present there are no validated means for significantly extending human lifespan.

Our ancestors were no less aware of ageing than we are, and were interested as to its cause and how it could be prevented. In the oldest known document about ageing, dating from 2,500 BC, an ancient Egyptian official named by Ptahhotep drew a gloomy picture:

> How hard and painful are the last days of an aged man. He grows weaker every day; his eyes become dim, his ears deaf, his strength fades; his heart knows peace no longer; his mouth falls silent and he can speak no word. The power of his mind lessens and today he cannot remember what yesterday was like. All his bones hurt.

It still strikes an all too familiar note.

There are important myths that we should bear in mind when we wish to extend longevity, notably that of Tithonus, who lived long but aged horribly. Efforts to disguise ageing by altering a person's appearance go back a long time, and Cleopatra certainly tried. There is now a multi-billion pound cosmetic and surgical industry devoted to limiting the physical ravages of getting old. The most common cosmetic surgery treatments are for face, breasts and fat; it is sexual attraction that seems to matter most. But some older people are joining in, even if the many facial creams for getting rid of wrinkles have only a minor effect in spite of all their claims.

Both ancient and current views about ageing are on the whole negative. There are of course significant differences in how different societies treat the old, as for instance in China or the US, but the way the old are viewed in our society is not as positive as we oldies would like. The old tend to be stereotyped as 'warm but incompetent'; and I do not like to be thought of as an 'old fogey'. As they age, people may become wiser, but at the same time are thought less competent in their jobs – even though, surprisingly, surveys do not bear this contention out. The old are now less welcome in public arenas such as politics. They are also often mocked in the press. The result is that the positive aspects of ageing are grossly neglected, and there is a failure to recognise that many of the elderly are quite happy with being old. As Mark Twain put it: 'Age is an issue of mind over matter. If you don't mind, it doesn't matter.'

There is, nevertheless, a lot of sympathy for the old and many people try to help them. In 1940 a group of individuals, as well as governmental and voluntary organisations, came together to form a committee to help old people and it soon gained national recognition. With the birth of the welfare state in the 1950s, government money became available to fund local work with older people and the committee became completely independent of government and took a new name – Age Concern. It became a national agent for schemes run by local groups, and drew attention to the plight of older workers who were unable to return to work because of long-term unemployment or redundancy. In spring 2010 Age Concern England joined together with Help the Aged to form Age UK, a new charity dedicated to improving the lives of older people.

Age UK combats ageism in all its forms, both social and as manifest in the treatment of the health of the old. Some of those responsible for medical care seem to see little point in spending much money and effort in keeping the old with serious illnesses alive. Age discrimination is present in many current societies – the compulsory UK retirement age being an obvious case – and when this occurs in a medical setting the results can be very serious and damaging.

But what really matters is how the old are treated. The worst cases, fortunately rare, are those where the old have been put to death when they were no longer thought to be of value to a society. Many of the old are lonely and poor. Most of the elderly want to stay in their own homes, and this can require support from others as well as money. Many of the old will end up having care at home, or living in a care home or nursing home. The money for this is partly provided by the state but there are complaints that it is inadequate, and many have to sell their homes. There are also problems with respect to professional care; many cases of incompetence and neglect are reported, even in hospitals. Dementia patients do especially badly.

The economic realities of an ageing society are only just beginning to impact on us. There are those who oppose extending human life partly because it is bad for the young, and because there are negative economic implications. A society in which the old greatly outnumber the young faces many challenges. Who will support all those elderly, and pay all their health costs? Countries like Japan and China face similar problems.

Ageing makes one think of death: how one should prepare for death, and how one should die. Most people want to die at home – I certainly do – but most people don't in fact do so. Do we die of old age? There is no good evidence that we can die of old age, and it is rarely put on a death certificate. In the US the use of these words alone is forbidden. Then there is the question of the best way for the old to die. Suicide is not uncommon, but euthanasia would be far preferable and the law preventing it must be changed; the old should have the right to choose how they die, while taking into account the pain loved ones feel for the dying.

A final surprise. Happiness, it seems, peaks at the age of 74 – or so Austrian and German scientists have concluded after asking 21,000 people how happy they were on a scale of 1 to 7. Teenagers registered around 5.5; in their 40s people reckoned they had less happiness, and those who were 74 rated themselves 5.9, the highest of the lot. This change in happiness was apparently most pronounced amongst British respondents; German men and women reported relatively stable levels of satisfaction throughout their lives. Dr Carlo Strenger, an Israeli psychologist, commented: 'If you make fruitful use of what you have discovered about yourself in the first half of your life, the second half can be the most fulfilling.'

So if you are told you are looking well, and are feeling happy, enjoy it for as long as you can.

2
Ageing

'The minute a man ceases to grow, no matter what his years, that minute he begins to be old' – William James

Our bodies change as we age. Looking stooped, for example, is a common sign of ageing. Most medieval pictures of the old show a bent back and a stick, and this continued into the twentieth century. In ancient Roman times, Virgil complained that 'all the best days of life slip away from us poor mortals first: illness and dreary old age and pain sneak up, and the fierceness of harsh death snatches us away.' Plutarch too had a gloomy image of old age, likening it to autumn. When children are asked how they can tell when people are growing old, they list physical attributes. Here we look at the major and minor physical health changes that are linked to ageing.

Ageing is not a disease, but is a multi-factorial process that leads to the progressive loss of functions. We are all too well aware of normal bodily changes as we age. We initially get a bit slower and then a little grey and bald, and then wrinkles come and memory goes. Cross-sectional studies of ageing tend to depict an essentially smooth and progressive decline of physiological function with increasing chronological age. However, although the young have high functional values and the very old low, between these limits values are widely scattered. There is no simple linear relation between age and functionality. When I meet some old friends whom I have not seen for some time I sometimes say, 'Shall we start at the top or the bottom?' We then tell about the pain in our foot, and work up the body to describe how our brain has declined.

One of the fairy tales collected by the Brothers Grimm in the early nineteeth century, 'The Old Hound', illustrates changes brought about by age:

> A hound who had served his master well for years, and had run down many a quarry in his time, began to lose his strength and speed owing to age. One day, when out hunting, his master startled a powerful wild boar and set the hound at him. The latter seized the beast by the ear, but his teeth were gone and he could not retain his hold; so the boar escaped. His master began to scold him severely, but the hound interrupted him with these words, 'My will is as strong as ever, master, but my body is old and feeble. You ought to honour me for what I have been instead of abusing me for what I am.'

Another of Grimms' fairy tales, 'The Duration of Life', collected from a peasant in his field in 1840, presents a pessimistic outcome but adds a playful teleological explanation:

> When God created the world he gave the ass, the dog, the monkey and man each a life-span of thirty years. The ass, knowing that his was to be a hard existence, asked for a shorter life. God had mercy and took away eighteen years. The dog and the monkey similarly thought their prescribed lives too long, and God reduced them respectively by twelve and ten years. Man, however, considered the thirty years assigned to him to be too brief, and he petitioned for a longer life. Accordingly, God gave him the years not wanted by the ass, the dog, and the monkey. Thus man lives seventy years. The first thirty are his human years, and they quickly disappear. Here he is healthy and happy; he works

> with pleasure, and enjoys his existence. The ass's eighteen years follow. Here one burden after the other is laid on him; he carries the grain that feeds others, and his faithful service is rewarded with kicks and blows. Then come the dog's twelve years, and he lies in the corner growling, no longer having teeth with which to bite. And when this time is past, the monkey's ten years conclude. Now man is weak headed and foolish; he does silly things and becomes a laughing stock for children.

There are few if any organs in our body that do not decline in their function with age, and many deaths are due to age-related illnesses. But not everything is bad news. A major study by ELSA (English Longitudinal Study of Ageing) in the UK is designed to find out about the health of the elderly, and participants are interviewed every two years. It is encouraging and impressive that 60 per cent of those aged 80-plus describe their health as good to excellent. But that does mean that 40 per cent have health problems. The study also found that while arthritis is age-related, joint pain and back pain were not, and were no more common among the elderly than the young.

The study looked at the proportion of people who remain free of certain diseases, including four eye diseases, seven cardiovascular diseases and six other physical diseases. Around half of those aged 50–54 still had none of those diseases, but only around one in ten of those aged 75–79. Wealth and education lead to longer physical functioning, possibly because both lead to better personal care. Money matters: people in the richest part of London live 17 years longer than those in the poorest parts. Individuals who are 50–59 years old and from the poorest fifth of the population are over ten times more likely to die earlier than their peers from the richest fifth. The poor are more likely to be unhealthy, despite a fairly even distribution in the quality of healthcare between different wealth groups.

In a different study, it was found that participants with a high IQ as a child were more likely to have better lung function at the age of 79. This could be because people with higher intelligence might respond more favourably to health messages about staying fit.

Disability and frailty are common problems for the elderly. Those who are ill experience ageing very differently from those who are well. There are 75-year-old joggers and 75-year-olds who are very frail. Frailty is a condition associated with ageing whose symptoms include weight loss, decreased muscle mass and strength, weakness, lack of energy and reduced motor performance. The condition seems to spring from a general weakening of the body, including the skeletal, muscular, blood and hormonal systems. The most commonly used measures of disability are reports of problems with the basic activities of daily living such as mobility, looking after oneself by preparing meals, shopping, managing money and taking medication. While disability indicates loss of function, frailty indicates instability and the risk of loss of function. The frail person is at increased risk of disability and death from minor external stresses. Frailty may also be identified by particular clinical consequences such as frequent falls, incontinence or confusion. In many cases a single factor, such as undetected cardiovascular disease, can be the reason why people become frail. Instead of having classic symptoms such as a heart attack or a stroke, people may have partly blocked blood vessels in the brain or the legs, the kidneys or the heart, which can result in exhaustion or mental confusion or weakness or a slow walking pace. It has been found that those people who had a positive outlook on life were significantly less likely to become frail.

As we age, our cells become less efficient and our bodies become less able to carry out their normal functions. Muscles lose strength, hearing and vision become less acute, reflex times slow down, lung capacity decreases, and the heart's ability to pump blood may be affected. In addition, the immune system weakens, making it less able to fight infection and disease. Heart pumping giving maximum oxygen consumption declines about 10 per cent every ten years in men, and in females a bit less; maximum breathing capacity declines about 40 per cent between ages 20 and 70; the brain shrinks and loses some cells; kidneys become less efficient and the bladder gets smaller; muscle mass decreases by about 20 per cent between 30 and 70 years, though exercise reduces this; and bone mineral is lost from age 35; sight may decline from 40 and hearing declines when older. These changes in our bodies with age are not due to ill health but are, alas, normal, and they can cause health problems.

Once adults reach 40, they start to lose just over 1 per cent of their muscle each year. This could be due to the body's failure to deliver nutrients and hormones to muscle because of poorer blood supply. Tendons, which connect muscles to bone, and ligaments, which hold joints together, become less elastic and are easier to tear. We also get slower in physical activities, as I know all too well. The good news is that one in five people aged between 65 and 74 are doing recommended levels of exercise. But physical labour can also have negative effects – lawyers and priests over 55 die at lower rates than blacksmiths and ironworkers, and at even lower rates when over 75. Mammalian muscles can regenerate, but in mice the old muscle regenerates poorly. Joining the muscles of old and young mice together resulted in the old muscle regenerating better, and the young a bit worse.

Men and women between 60 and 96 years of age who suffered from loss of body mass and strength, and who did a moderate amount of strength training twice a week, had a significant increase in muscle after eight to twelve weeks. Long-term physical activity postpones disability and sustains independence, even for the chronically ill. Regular physical activity can also help to prevent some important conditions in the elderly that may lead to disability including osteoporosis, type 2 diabetes, cardiovascular disease, anxiety and depression. It can also reduce the risk of falls and therefore subsequent fractures. A goal to work towards is 30 minutes of at least moderately intense physical activity on at least five days of the week. Joan Bakewell, 77, of whom more later, told me, 'I am not as able as I was and do have some aches and pains, but I have a fetish for staying fit and have done pilates exercises twice a week for fifteen years. I did not want my posture to go, as I noticed I was beginning to stoop.'

Most human deaths are attributable to an age-related disease and so, not surprisingly, becoming a centenarian is associated with having avoided common diseases until advanced in age. A period of not having good health for the elderly will usually involve some seven years for men, and ten for women. Coronary heart disease, stroke, cancer, osteoporosis, diabetes and dementia are just some of the conditions that more and more people will be battling with in later years. The figures will continue to go up as people live longer than ever before. Obesity increases the likelihood of death from all causes, particularly coronary artery disease and stroke. In addition to these diseases, obese patients suffer an increased incidence of arthritis. The conditions associated with obesity are also associated with ageing. The proportion of intra-abdominal fat, which is related to increased morbidity and mortality, progressively increases with age. Targeting weight loss in the elderly can therefore reduce morbidity from cardiovascular risk factors, and also arthritis.

The leading causes of death in the United States have changed dramatically. In 1900 the top three causes of death were all related to infectious diseases, but by 2009 the leading causes of death for all ages were diseases of the heart and cancer, which together account for 50 per cent. The top four causes of death for persons aged 65 and older – diseases of the heart, cancer, brain dysfunctions related to disease of the blood vessels supplying the brain, and respiratory diseases – were the same as for all ages. In the UK the main causes of death are heart disease and stroke. There is an elusive distinction between the effects of ageing, and having a disease when old, but it is generally accepted that fundamental to ageing is an increasing vulnerability to diseases such as heart disease.

Coronary heart disease is the leading killer of older people and half of all heart attack victims are over 65. Participating in light to moderate physical activities significantly decreases mortality rates in elderly patients. Heart disease is the result of the heart's blood supply being blocked or interrupted by a build-up of fatty substances in the coronary arteries. Over time, the walls of arteries can become furred up with fatty deposits, a process known as atherosclerosis. When the coronary arteries become narrow due to a build-up of these fatty deposits, the blood supply to the heart will be restricted. A heart attack occurs when blood flow to an area of heart muscle is completely blocked. This prevents oxygen-rich blood from reaching heart muscle and so causes it to die, and then circulation of the blood fails. Without quick treatment, a heart attack leads to serious problems and often death. While men have markedly higher rates of coronary heart disease in middle age than do women, women's rates of coronary disease begin to rise sharply after menopause.

Failure of the blood supply can also cause strokes. A stroke occurs when the blood supply to the brain is disrupted in some way and the brain cells are deprived of oxygen and other nutrients, causing some cells to become damaged and others to die. The effects of a stroke depend on where the brain was injured, as well as how much damage occurred. The after-effects of stroke are very varied and depend on how and where nerve cells die. Most of the damage caused to the brain is the result of dying nerve cells releasing toxins that damage even more of the brain. A stroke can impact on any number of areas including the ability to move, see, remember, speak, reason, and read and write. Every year about 150,000 people in the UK have a stroke – that's one person every three and half minutes, every day. A study of elderly stroke victims in a working-class region of London found that many regarded it as a normal crisis in their lives. Although strokes can happen at any age, the vast majority occur in people over 65 years old. However, a new study shows more and more Americans suffering from stroke earlier in life than ever before, indicating that stroke is no longer just an affliction of old age.

Cancer is fundamentally a disease of the elderly, the incidence of cancer in those over 65 being ten times greater than in those younger. Nearly three quarters of cases are diagnosed in people aged 60 and over, and more than a third of cases in people aged 75 and over. More than three quarters of cancer deaths occur in people aged 65 years and over. Although there is a higher number of cancer deaths in the over 65s, cancer causes a greater proportion of deaths in younger people. Among elderly men, cancers of the prostrate and colon are the most common, while for women it is breast cancer. About 80 per cent of all breast cancers occur in women aged over 50.

The relationship between cancer and ageing is quite complex. Genomic instability, DNA damage, is a hallmark of most cancers, and is also a hallmark of ageing, as we will see. There have to be a number of changes to the genes within a cell before it turns into a cancer cell. It then takes time for further changes to occur for the cells to become malignant. The relationship to ageing is probably related to the increased time cells have to develop abnormalities that increase the risk of cancer, and also the increase in time that they are exposed to a cancer-inducing environment. A surprising statistic is that the potential gain in life expectancy which could result from the complete elimination of mortality from cancer in the US would not exceed three years if one were to consider cancer independently of other causes of death.

Cells can respond to new situations by increasing their growth in ways that do not necessarily lead to cancer, but have other effects. The prostate in older men is all too good an example. This is due to increased proliferation of the epithelial cells and fibroblasts, and an increase in size of the smooth muscle. This results in increased urination as the prostate protrudes into the bladder. Why this should occur with age is just not known, and it all too often can develop into a tumour. But not all abnormal increases in growth are cancerous, thankfully.

Decline of the immune system with age is serious. In the elderly, many alterations of both innate and acquired immunity have been described. This process is responsible for increased susceptibility to infectious diseases such as flu and pneumonia, as well as being at the root of the biological mechanisms responsible for inflammatory age-related diseases. In the USA 90 per cent of deaths from flu and pneumonia are in people aged over 65 years.

Osteoarthritis is a type of arthritis and some 8 million people in the UK suffer from it. It affects mainly older people starting at around age 45, and is caused by the breakdown and eventual loss of the cartilage that serves as a cushion between the bones of the joints. Loss of the cartilage cushion causes friction between the bones, leading to pain and limitation of joint mobility. People with osteoarthritis usually have joint pain and limited movement. Unlike some other forms of arthritis (there are over a hundred types), osteoarthritis affects only joints and not internal organs. More than half of the population aged 65 or older would show X-ray evidence of osteoarthritis in at least one joint. Both men and women have the disease. It is due to a combination of factors, including being overweight, the ageing process, joint injury, and stresses on the joints from certain jobs and sports activities.

Osteoporosis is a silent disease in which bones become extremely fragile. The bone mineral density is reduced, and bone structure is disrupted. If left untreated, it can progress painlessly until a bone breaks, typically in the hip, spine or wrist, and these breaks are extremely painful and can take a long time to heal. It is claimed that one in every two women and one in every four men over 50 will break a bone due to osteoporosis. Gout, however is not age-related.

Falls are amongst the most common and serious problems facing the elderly. They can lead to death, reduced function and admission to a nursing home. About half of the over-65s have a fall each year and the numbers increase with age. Cognitive impairment and dementia increase the risk of falls and the risk of fractures to bones is greater. Balance problems and dizziness are considerably more common the older the person – three out of five women aged 80 and over experienced one or both of these at least sometimes, compared with only one out of five women in their 50s. Repeated falls can lead to the need for long-term care. Difficulties with walking and climbing stairs are all too common amongst those over 75. More than 2,300 older people fall every day and 80,000 of those who have fallen during the year are afraid to leave their homes. Exercise programmes can help in preventing falls.

Older people have an abnormal gait when walking, and this is evident in about one third of those over 65. The characteristics include slow speed, short step length and large variability, narrow stride width, and stepping frequency. Uneven pavements – in London one in five pavements is in a poor state of repair – can be hazardous. Elderly people could one day be relying on a bodysuit, rather than a Zimmer frame, for support. Scientists in Japan are developing the 'Michelin gran' Lycra suit which is covered in pairs of inflatable 'muscles' which assist the wearer's real muscles. When they inflate, they help the wearer move their limbs with more strength and stability.

Type 2 diabetes is most often seen in older adults, with half of all cases diagnosed in people of 55 and over. It is by far the most common form of diabetes, and occurs when the body produces insulin but the cells no longer respond by allowing entry of glucose, especially in muscle, fat and liver cells. About 90 per cent of patients who develop type 2 diabetes are obese. Both being overweight and lack of mobility promote the disease.

There are many false ideas about the sexual activity of the elderly. It has been said that sex among the old is a well-kept secret because the young would not believe it if they were told. In fact there is little evidence for a significant age-related decline. The level of sexual interest and activity among people over the age of 65 is as diverse as the individuals who make up that population. A recent survey in the US of married men and women showed that nearly 90 per cent of married men and women in the 60–64 age range are sexually active. Those numbers drop with advancing years, and a bit less than one third of men and women over the age of 80 are still sexually active. A recent survey found that men enjoy five extra years of an active sex life – up until the age of 70 – compared to women, who were less likely to make love after the age of 65, mainly because they had married older men and their partner had died. Those who are healthy in older age are twice as likely to enjoy a high libido and more likely to have regular sex, that is once or more a week. Even so the male sexual response tends to slow down with age.

There is, however, significant erectile dysfunction and impotence in those over 70 – about one third suffer from this. Physical illness is a common reason for ending sexual activity, and medications such as antidepressants, statins and benzodiazepines can have similar effects. By the age of 65, about 15 to 25 per cent of men have this problem at least one out of every four times they are having sex. This may also happen in men with heart disease, high blood pressure or diabetes, either because of the disease or the medicines used to treat it. A man may also find it takes longer to get an erection and his erection may not be as firm or as large as it used to be. The amount of ejaculate may be smaller. The loss of erection after orgasm may happen more quickly, or it may take longer before an erection is again possible. Some men may find they need more foreplay.

It is well-documented that older women experience fewer sexual problems than men as they age. Most healthy women can expect unimpaired sexual activity to the end of their lives, if that was their pattern earlier. Normal changes in the older woman include a decrease in length, width and elasticity of the vagina. Recent studies, however, indicate that the older woman has no physical limitation in her capacity to achieve and enjoy orgasm. But there are some limitations. The decline in the female hormone, oestrogen, which occurs after menopause, can result in decreased vaginal lubrication. The loss of lubrication can often result in painful intercourse, but fortunately this condition can be easily treated with creams. Diana Athill, an award-winning writer over 90, wrote that she has given up sex and says she doesn't miss it. 'It's like not being able to drink wine – at first I thought that was a terrible detriment, but once you can't drink something, and it makes you ill if you do, you don't mind giving it up . . . One reads from time to time absolutely obscene articles about senile sex – about how if you really go on trying hard enough, using all kinds of ointments, it can work, but for God's sake! It's supposed to be fun! If you need a cupboard full of Vaseline, you might just as well stop.'

Changes in our skin giving rise to wrinkles in the face are a major sign of ageing. A study of the basis of their formation using gene technology claims to have identified more than a thousand genes and their proteins that are involved. One pathway results in the loss of water, while another is the breakdown of collagen, a molecule that gives the skin strength, and a third is damage from sunlight. With ageing, the outer skin layer thins, even though the number of cell layers remains unchanged. The number of pigment-containing cells decreases, but the remaining ones may increase in size – age spots – in sun-exposed areas.

Ageing skin is thus thinner, pale, and more translucent. Changes in the connective tissue reduce the skin's strength and elasticity, especially in sun-exposed areas. It produces the leathery, weather-beaten appearance common to those who spend a large amount of time outdoors. The blood vessels under the skin become more fragile, which in turn leads to easy bruising, but most bruises go away without treatment. The skin glands produce less oil with age and while men experience a minimal decrease, usually after the age of 80, women gradually produce less oil after menopause. This can make it harder to keep the skin moist, resulting in dryness and itchiness. The fat layer, which provides insulation and padding, thins, and this increases the risk of skin injury and reduces the ability to maintain body temperature in cold weather. The sweat glands produce less sweat and this makes it harder to keep cool in hot weather, and so increases the risk of becoming overheated. Growths such as warts and other blemishes are more common. Skin-healing can sometimes fail completely in later life, causing a wound to become chronic. Statistics suggest that at least one in twenty people over the age of 65 have a non-healing skin wound. At particular risk are older people with diabetes, and a distressed emotional state can impair healing even further.

Eye diseases like cataracts, glaucoma and macular degeneration cause loss of vision and are major problems in old age. Cataracts are areas that distort light as it passes through the lens of the eye. As we age, protein in the lens of our eyes can clump together and cloud the lens. Glaucoma is an eye condition in which the fluid pressure mounts inside the eye; this pressure can harm the optic nerve. It is often hereditary and worsens with age. Both cataracts and glaucoma can be treated. Macular degeneration is a disease that causes progressive damage to the central part of the retina that allows us to see fine details. In the USA 30 per cent of patients aged 75 to 85 will have macular degeneration There are genes which increase the risk, and diabetes and high blood pressure also increase the risk of eye problems. There is also presbyopia, the typical 'long-sightedness' of middle age, in which people find it more and more difficult to read small print. To begin with they cope by holding the reading material further and further away.

About one-third of Americans between the ages of 65 and 74 have hearing problems and about half the people who are 85 and older have hearing loss. In the UK there are more than 6 million deaf and hard-of-hearing people aged over 60. Tiny hairs inside your ear help you hear as they pick up sound waves and change them into the nerve signals that the brain interprets as sound. Hearing loss occurs when the tiny hairs inside the ear are damaged or die. The hair cells do not regrow, so most hearing loss is permanent. Another type of hearing loss results from damage to other parts of the inner ear. Tinnitus can occur with many forms of hearing loss, including those that sometimes come with ageing. People with tinnitus may hear a ringing or some other noise inside their ears. The good news is that hearing deterioration tends to halt at around the age of 70.

We visit the dentist more as we age. Teeth provide a good model of wear and tear as an ageing process. With age our teeth appear to get darken, due to changes in the dentine beneath the surface enamel. The enamel itself can become worn down from years of chewing, and this causes teeth to become more sensitive. The teeth themselves also become dryer and more brittle, which makes them more likely to break or crack during normal chewing, and old fillings may start to fracture. Gums can start to recede, especially if there is a periodontal disease or they have been subject to too-forceful brushing. Receding gums can increase the risk of tooth decay.

Approximately a quarter of men begin balding by the age of 30 and two-thirds begin balding by age 60. One hypothesis suggests baldness evolved in males as a signal of ageing and social maturity, showing that aggression and risk-taking decrease. This could enhance their ability to raise offspring to adulthood. Most of the hairs on a person's head are in an active growth phase, which may last anywhere from two to seven years. At the end of this stage the cells causing growth, which are stem cells, die, and the hair falls out. The average scalp contains about 100,000 hairs and roughly 100 hairs are lost every day. Baldness is due to the failure to replace lost hairs and has a genetic basis, but stress can also cause hair loss.

The greying of hair can also occur at quite a young age but commonly begins in the mid-thirties. The blackness of hair is due to special dark-pigment-producing cells, melanocytes, entering the growing hair, and greying is due to their absence or failure to produce dark pigment. For unknown reasons, hair stem cells have a much greater longevity than the melanocyte stem cells, so greying can occur before baldness. Stress hormones may impact the survival and activity of melanocytes, but no clear link has been found between stress and grey hair. Genes can affect both baldness and greying, and twin studies showed that female greying is genetic.

If all this information sounds depressingly negative, it may be worth remembering 'The Old Man and the Three Young Men', a verse fable by La Fontaine, one of the most widely read French poets of the seventeenth century, which gives strength to the old:

> An Old Man, planting a tree, was met
> By three joyous youths of the village near,
> Who cried, 'It is dotage a tree to set

At your years, sir, for it will not bear,
Unless you reach Methuselah's age:
To build a tomb were much more sage;'

But all three youths die from accidents.

The Old Sage, then,
Weeping for the three Young Men,
Upon their tomb wrote what I tell.

And in most areas of physical activity, not least sport, there are positive and encouraging examples of physical ability with increasing age. Do those in various sports truly begin an unstoppable physical decline in their twenties, or is this more of a question of reduced motivation? 'How old would you be if you didn't know how old you were?' asked Leroy Paige while playing major league baseball as a pitcher into his late fifties. Ageing is relative – the 60-year-old tennis player may be in better shape than the 20-year-old couch potato. I still regularly play both doubles and singles.

Not long ago it was 'time to hang it up' when an athlete hit 40, and one was definitely regarded as being over the hill at 50. Now it's more like 75. This refers to those who are training and trying to improve their performance, not the average couch-lounger. Some have kept training as they got older, and others have started competing in middle age. One study showed year-to-year performance declining after 50, but at a rate that was barely noticeable until about 75, when the decline became undeniable. Male and female senior athletes' performance declined approximately about 4 per cent per year over 35 years of competition – slowly from age 50 to 75 and dramatically after 75. The decline in the sprint was greater than in endurance for women, especially after 75. Marathon runners decline about 2 per cent a year between 30 and 40, 8 per cent between 40 and 50, then 13 per cent 50 to 60, and finally there is a 14 per cent decline between 60 and 70. But in 2003 the Canadian long-distance runner Ed Whitlock ran a marathon in under three hours at the age of 73.

There are other amazing exceptions to decline of athletic ability with age. Luciano Acquarone was a marathon world-record holder at 59 years with a time of 2:39, and a 60-year-old runner from Japan who ran it in 2:36. There is a 73-year-old lady who tries to swim, cycle and run three times a week as it makes her feel young and live longer, and she will even enter the Hawaii Ironman triathlon. A great example of strength with age is Ranulph Fiennes, who climbed Everest at the age of 65, and six years earlier ran seven marathons in seven continents on seven successive days. Ruth Frith at 100 set a new world record in her age category with her shot put of just over 4 metres – longer than the throws of competitors decades younger – at the World Masters Games in Sydney in 2009. She trains five days a week, pressing 80lb weights

'Old soccer players never die, they just achieve their final goal.' Roger Milla was 42 years old when he played in the 1994 World Cup for the Cameroon national team. He achieved international stardom at 38 years old, an age at which most footballers have retired, by scoring four goals at the 1990 World Cup. Many football players are in their prime in their 30s.

Snooker requires both mental and physical skills. Ray Reardon, who claimed six world snooker titles, reached the 1982 world final at the age of 49, only to lose to an inspired Alex Higgins. One of his greatest moments came in 1988 when, aged 56, he thrashed Steve Davis, winner of five world crowns. Golfers too can continue through the years. Tom Watson at 59 only just lost the 2009 British Open, and there was a great deal of positive news coverage that someone of his age could be so successful. One old golfer was still driving himself to the course at 101, and only recently moved into assisted-care living at Arizona Grand. At 100, he could still break 40 for nine holes, and he didn't start riding in a cart until a hip problem caught up with him at 98.

Declining health and capability that accompanies age may be made worse by prolonged exposure to an unhealthy environment and lifestyle. Older people rather consistently rate their health as good, despite evidence suggesting that they are more likely to suffer from a variety of health problems. Many older people take a holistic view of what 'health' means, including wellbeing and social factors, and in general take a positive view. Older people who accept negative images of ageing are more likely to suffer with health problems – they are also, because of their negative attitude to ageing, more likely to attribute their problems to the ageing process, and therefore do not seek the necessary medical assistance. Some older people may also minimise their health problems as a deliberate method of denying negative age-related stereotypes. In addition, some older people are reluctant to visit medical professionals because they feel, unwisely, more comfortable trusting their own common sense.

Heart-felt perils await people who hold disapproving attitudes about the elderly, a new study suggests. Young and middle-aged adults who endorse negative stereotypes about older people display high rates of strokes, heart attacks and other serious heart problems later in life, compared with ageing peers who view the elderly in generally positive ways. Yale University psychologist Becca Levy found that those who viewed ageing as a positive experience lived an average of seven years longer. This means that a positive image had a greater impact than not smoking or maintaining a healthy weight. Levy says that patronising attitudes and 'elderspeak' – speaking to elders as if they were children – can affect their competence and lifespan. There are claims that optimism and coping styles are more important to successful ageing than physical health. But do keep doing exercise. Also note that there is a little evidence that being bored can result in heart problems and an earlier death.

We will look later a the possible cellular basis of the changes in our bodies as we age, but now turn to mental changes and illnesses associated with ageing.

3
Forgetting

'First you forget names, then you forget faces, then you forget to pull your zipper up, then you forget to pull your zipper down'
– Leo Rosenberg

Almost all of us who are ageing must admit that our minds deteriorate in some ways as we age. Memory is the most obvious example: it is all too common for simple things, particularly names, to be forgotten, and then often they come back unexpectedly a little while later. In fact speed of thought and spatial visualisation begin to decline in our twenties and our mental abilities peak at about 22 years old, before beginning to deteriorate just five years later. Memory has been shown to decline from an average age of around 40. But abilities based on accumulated knowledge, as shown in performance on tests of vocabulary or general information, increase until the age of 60. George Burns pointed out: 'By the time you're eighty years old you've learned everything. You only have to remember it.' And though our mental processes do become less efficient, we can become more efficient in solving complex moral and social problems.

The Ancient Greek dramatist Sophocles was hardly positive about old age: 'When a man is old the light of his reason goes out, action becomes useless and he has unmeaning cares.' His sons appeared to agree: because of his absorption in literary work, Sophocles was thought to be neglecting his business affairs, and his sons hauled him into court in order to secure a verdict removing him from the control of his property on the ground of imbecility. Then the old man read to the jury his play *Oedipus at Colonus*, which he had written at the age of over 80, and inquired: 'Does that poem seem to you to be the work of an imbecile?' When he had finished, he was acquitted by the verdict of the jury.

The Roman orator Cicero wrote: 'Old men retain their mental faculties, provided their interest and application continue; and this is true, not only of men in exalted public station, but likewise of those in the quiet of private life.' It is impressive to see how similar his views about how to age best are to current ones – keep mentally and physically active. To the objection that memory begins to fail in old age, he replies: 'No doubt it does, if you don't keep it in trim, or if you happen to have been born a trifle dull. I have never heard of any old man forgetting where he had buried his treasure: the old remember what is of real concern to them: their days in court, their debts, and their debtors.'

Those reaching 70 with some 14 years ahead will, on average, spend nearly two years in moderate or severe cognitive impairment. There is thus great concern, even fear, about the increase in dementia with age, especially Alzheimer's disease. The Disconnected Mind team is conducting an extensive analysis on the contribution of lifelong demographic factors to the rate of brain ageing. Brain function at age 11 is the biggest indicator of brain function in later life; clever children make clever and mentally competent old adults.

There is not a significant loss of nerve cells with age,

though several populations of nerve cells are lost in the old. As people age, peripheral nerves may conduct impulses more slowly, resulting in decreased sensation, slower reflexes, and often some clumsiness. Nerve conduction slows because myelin sheaths, the layers of cells around nerves that speed conduction of impulses, degenerate. Shrinkage of the brain is due to fluid loss as well as reduced branching of extensions from nerves. There is also a reduction in the size of the cell bodies and the accumulation of granular pigment and filament tangles and abnormal small structures. In about one third of old brains, there are abnormal deposits of amyloid, which are proteins that form insoluble fibrous protein aggregates and are common in the brains of those with Alzheimer's disease, and this leads to damage of nerve cells. Late onset schizophrenia is rare.

Studies have revealed that separate brain regions that are involved in higher-order cognitive functions show less-coordinated activation with ageing. This reduced coordination of brain activity is associated with poor performance in several cognitive domains. Although neuronal loss is minimal in most regions of the normal ageing brain, changes in the connections between ageing neurons may contribute to altered brain function. More than 150 genes have been found to undergo age-dependent expression changes in the brain; some of these are more active with age in mice but less so with age in humans. The function of these genes and how they are turned on and off is not yet understood. Studies on mice have identified memory disturbances in the ageing brain as being due to certain genes associated with memory being turned off. There are also lower concentrations of neurotransmitters like dopamine, which has important roles in behaviour, cognition and voluntary movement. In the human brain, declining mitochondrial function may selectively affect neuronal populations with large energy demands, such as the neurons that degenerate in Alzheimer's disease.

With mental activities, there is usually a much less dramatic decline with age than with physical activities. It varies a great deal, but the old can still be very productive. Politicians can continue to be active until rather old, some might say too old. Roman emperors ruled until they were of extreme old age. Augustus, who lived until the age of 76, remained in office until his death and didn't stop visiting the Senate on a regular basis until he was 74. Winston Churchill was still prime minister at the age of 80.

Scientists usually do their best work when young, but there are important exceptions, such as Galileo writing *Dialogues Concerning the New Sciences* when he was 72. The physicist Max Planck wrote: 'Scientific theories don't change because old scientists change their minds, they change because old scientists die.' It has been found that science professors in their 50s and 60s published almost twice as many papers each year as those in their early 30s. When he was 80, André Gide said that he did not detect any weakening of his intellectual powers but did not know what to turn them to.

Writers, painters and sculptors do not lose their skills with age, and many have produced their finest works in the last fifteen or so years of a long life. At 97, Enrico Paoli, an Italian chess master, was the strongest active nonagenarian chess player in the world. He learnt chess when he was nine and started playing tournaments at 26. He won his last Italian championship title at the age of

60. Paoli was playing master-level chess at 96 – in 2003 he played the international tournaments. Jose Raul Capablanca, the 'Mozart of chess', regarded Emanuel Lasker, who was world chess champion for 27 years, as the most dangerous player in the world in a single game, even as the latter neared 70. No other contemporary, he thought, surpassed him in his ability to evaluate a position and find the correct strategy.

The decline of memory with age depends on the specific nature of the memory, as there are different types. For example a patient with a particular brain damage cannot recollect personal experiences, but can learn new motor skills and lists of words. This is an implicit memory and involves recall of motor and academic skills without conscious awareness of previous experiences. It is distinct from explicit memory, which involves recall of previous experiences and information. Explicit or episodic memory involves the memory of autobiographical events, including recent events, such as times, places and associated emotions and is the most common memory loss with age. As the length and complexity of sentences increases, older adults have more difficulty understanding and recalling them. Yet factual knowledge does not decrease with age, though spatial memory, such as the layout of a museum recently visited, does decline.

It is common with old age to forget names of people or to lose a particular word – even though it is on the tip of the tongue. Usually the name or word is recalled later when one is thinking about something quite different. I have lost names, so too have many of my friends. I have also forgotten the faces of people whom I know quite well and have to ask them who they are when they greet me, and then I recall who they are. It is a bit embarrassing not to recall the name of someone you know when you meet them and need to introduce them to someone else. One also loses common objects, or as Edward Grey put it: 'I am getting to an age when I can only enjoy the last sport left. It is called hunting for your spectacles.'

Jonathan Swift, the author of *Gulliver's Travels*, describes a familiar experience:

> That old vertigo in his head
> Will never leave him, till he's dead:
> Besides, his memory decays,
> He recollects not what he says;
> He cannot call his friends to mind:
> He forgets the place where he last dined:
> Plies you with stories o'er and o'er,
> He told them fifty times before.

Though there is no significant loss of knowledge with age, the elderly do not encode information into long-term memory as efficiently as the young. Forgetting to do something as one ages is common and worrying. Around 60 per cent of participants in a study of those aged 75 and older forgot to perform an action that they had previously been requested to carry out. A typical example of loss of a recent memory is the case of a distinguished but ageing TV presenter who went out to dinner on a Friday night. When he rang the hostess's bell there was a delay, then she put her head out of an upper window and said hello. 'Have I come on the wrong night?' he asked. 'No,' she replied, 'it was last Friday and you were here.' All too familiar.

Complaints about memory are the most frequent cause for seeking medical advice about dementia. This is the result of episodic memory going wrong and leads to forgetting personal and family events and appointments; losing items round the house; repetitive questioning; inability to follow plots on TV or in films; forgetting past events and news items; and getting lost. The elderly have many more memories for events that occurred in adolescence and early adulthood than in midlife. Very few elderly show improved cognitive functioning in the evening, and unlike the young, their performance gets worse through the day. The herbal treatment ginkgo, used by the Chinese for thousands of years to overcome loss of memory in the old, has been shown to be totally ineffective. All the fuss over fish oil as a key brain food may be unjustified. A two-year study found there is no evidence that the supplements offer benefits for brain function in older people, contradicting previous surveys on the wonders of omega-3 fatty acids. But physical fitness contributed to more than 3 per cent of the differences in cognitive ability in old age after accounting for a participant's test scores at age 11.

About half of all lifetime cases of mental illness begin by age 14 but the chance of developing a mental disability increases as we age. The most common and serious one is dementia, an overall impairment in cognitive functioning sufficient to affect everyday activities, and there may be depression, hallucinations and delusions. The term dementia was introduced by Philippe Pinel in Paris in 1801; he also introduced the idea that people suffering from it should be treated with kindness – many patients at that time had actually been kept in chains – and he called this new principle 'the moral treatment of insanity'. One of Pinel's students, Dominique Esquirol, gave a very detailed and still valid description of dementia, pointing out, for example, that sufferers entertain perfect indifference to objects that were once most dear, and this includes relatives. They also often have a ridiculous passion. He carried out autopsies and noted abnormal convolutions in the brains of patients, but microscopic examinations of such brains had to wait for work on Alzheimer's.

While dementia is rare before 60, it increases with age and is present in 5 per cent of the over 65s and in 20 per cent of the over 80s. Dementia isn't a specific disease, rather it describes a group of symptoms affecting intellectual and social abilities severely enough to interfere with daily functioning. Different types of dementia exist, depending on the cause. Low education and diabetes can contribute, and mental and physical activity help prevent it. Alzheimer's disease is the most common form of this disease, and is the cause in two thirds of cases of dementia. It increases the risk of dying by two to five times, and for those over 85 accounts for one third of deaths, though this is rarely on the death certificate. Staff who look after Alzheimer's patients need to recognise it can be a terminal illness.

Dementia and normal ageing may be on a continuum. Memory loss generally occurs in dementia, but memory loss alone doesn't mean there is dementia. It can be difficult to distinguish between onset of Alzheimer's and ordinary ageing relapses of memory and cognition. Dementia has many causes and some dementias, such as Alzheimer's disease, occur on their own, not as a result of another disease. Some dementias, such as those caused by a reaction to medications or an infection, are reversible with treatment. Recent research shows that up to 80 per cent of people diagnosed with mild mental or cognitive impairment go on to develop much more debilitating dementia within just six years. As well as the personal devastation caused by cognitive decline, it is also the single biggest reason why older people lose independence and require 24-hour care.

There are some 820,000 people in Britain with dementia, about half of whom of whom suffer from Alzheimer's disease. Generally Alzheimer's is diagnosed in people over 65 years of age, although the less-prevalent early-onset Alzheimer's can occur much earlier. The disease in those in the 30 to 40 age group is rare, and even rarer are cases due to a genetic defect that results in it occurring in those as young as 16. There are estimates that there are over 20,000 younger people with dementia in the UK. The Alzheimer's Society estimates that the disease costs Britain £17 billion a year. Families caring for patients save the government £6 billion a year. It is expected that a further one million people will develop dementia in the next 30 years, and it is projected that expenditure on long-term care services for older people with dementia is set to increase from about £4.6 billion in 1998 to £10.9 billion in 2031. It is claimed that 5.3 million Americans are living with the disease; a new case develops every 70 seconds. An estimated 27 million people worldwide had Alzheimer's in 2006; this number may quadruple by 2050.

Dementia causes problems with at least two brain functions, memory loss along with impaired judgement and language. Dementia can make an individual not just confused and unable to remember people and names, but also experience changes in personality and social behaviour. Some causes of dementia are treatable and even reversible, but early diagnosis is important so that treatment can begin before symptoms worsen. If the diagnosis is a dementia that will progressively deteriorate over time, such as Alzheimer's disease, early diagnosis also gives a person time to plan for the future while he or she can still participate in making decisions. A quite simple test has been developed for detecting early stages of Alzheimer's that would require further investigation. It is a two-page questionnaire, and has ten tests that include remembering a phrase, doing sums, identifying parts of a man's suit, and drawing the time on a blank clock face. A quite different early diagnosis may be based on examining the fluid in the spine for the proteins that are responsible for the disease.

The disease is named after Alois Alzheimer, who was born in 1864 in Germany and studied medicine. He then worked at the Municipal Mental Asylum in Frankfurt, and later moved with Emil Kraepelin to the Max Planck Institute in Munich. He died in 1915. His first patient with the disease that was eventually to carry his name was Auguste D., a 51-year-old woman. She presented with jealousy of her husband, paranoia, memory impairment and, towards the end of her life, loud screaming. On 3 November 1906 Alzheimer presented the results obtained from studying the structure of the cellular organisation of her brain which included several fibrils – amyloid protein – and numerous small abnormal foci due to cell death. It was his mentor Emil Kraepelin who gave the disease his name in the eighth edition of his textbook in 1910.

At the early stages of Alzheimer's disease the most commonly recognised symptom is memory loss, such as difficulty in remembering recently learned facts. Typically there is misplacing and losing objects and repeatedly asking the same questions. The sufferer has difficulty in finding words to complete a sentence and comprehension is poor, as is doing complex motor tasks. There are also non-cognitive symptoms – delusions, depression and anxiety, and verbal and physical aggression. Patients can be very difficult. One patient hit his wife, and did not understand the difference between night and day – he could go to bed and get up thirty times in a night. Median survival is about five years. Apathy can be observed at an early stage, and remains a most persistent symptom. Despite the loss of verbal language abilities, patients can often understand and return emotional signals. A remarkable finding is that facial asymmetry in men results in an increased mental decline in the years before death.

Although it may not be easy, people with Alzheimer's disease can live quite full and productive lives. Taking time to prepare for the challenges that come as the disease progresses will ease the difficult transitions. Well-known personalities such as Bernard Levin and Iris Murdoch had Alzheimer's, and Terry Pratchett has a special form. He has problems recognising visual signals and can take minutes to tie a tie, but his talking is fluent, though his reading can be confused. Watching him and others with Alzheimer's on a BBC TV programme, there was nothing in the way they talked or behaved which gave any indication of abnormality. But they could not remember what had happened recently, and one could not copy a geometrical design. Pratchett has said he 'would eat a dead mole's arse' if it would cure him. He wants to be able to choose when to die. The TV journalist John Suchet has described the dementia of his wife, which started when she was 61. She now will pile dirty plates on top of clean ones, and repeatedly flush the toilet when not using it. Her memory is poor. Husbands and wives are six times more likely to get dementia if their spouse has it. This could be related to the stress of caring or living with someone in that condition.

Andrea Gilleas describes in her book *Keepers* the severe problems of looking after a mother-in-law with Alzheimer's. A recent play, *Really Old, Like Forty Five* by Tamsin Oglesby, takes place when Alzheimer's has reached epidemic proportions. A family are trying to deal with an elderly lady member suffering from the disease. There are, unusually, comic elements like robot nurses, midway between cat and carer, and bizarre plans to help by government officials.

Some cases of early-onset Alzheimer's may be caused by a number of different gene mutations. These mutations cause the formation of abnormal amyloid, fibrous protein aggregates in the brain. One of the ways proteins can have severe negative effects is by becoming an amyloid. Many different proteins can do this, developing sticky elements which enable them to stick together and form sometimes deadly fibres. The amyloid appears to bind to the neurones with a prion protein that is associated with BSE and Creutzfeldt-Jacob disease. The amyloid accumulation interacts with a tau protein which gets into nerve cells and causes the formation of aggregates of particles, leading to the tangles which cause the death of nerve cells. Recent studies correlated levels of the tau and amyloid proteins in the cerebrospinal fluid with changes in cognition over time, and found that changes in these two protein levels may signal the onset of mild Alzheimer's. This is a significant step forward in developing a test to help diagnose the early stages of Alzheimer's disease.

One predisposing genetic risk factor is related to the APOE genes that code for proteins that help carry cholesterol in the bloodstream. APOE comes in several different forms; APOE4 occurs in about 40 per cent of all people who develop late-onset Alzheimer's and is present in about 25 to 30 per cent of the population. People with Alzheimer's are more likely to have an APOE4 gene; however, many people with Alzheimer's do not have it. There is concern that a genetic test for this gene could lead to excessive anxiety. This is particularly true of individuals whose parents have suffered from the disease. A study of such individuals tested for the APOE gene did not find any serious anxiety in those who tested positive, but none suffered from anxiety or depression prior to the test. In addition, they were given counselling and followed up for a year. A surprising feature of the gene is that those who have it are more intelligent.

There is no cure for Alzheimer's, but a few drugs such as Aricept can improve memory a bit, and have general benefits including improving alertness and motivation for those in the early stages. It may take some months for there to be a noticeable improvement or slowing down of memory loss. Claims that the anti-histamine drug Dimedon had positive effects have now been shown to be wrong. Non-drug treatments include reality orientation with clocks, boards and newspapers; reminiscence therapy recalling past events like marriage; cognitive stimulation therapy like physical and mental games; and music —therapy. There is evidence that exercise and a diet rich in fruit and vegetables lowers the risk of getting Alzheimer's. A good education also lowers the risk.

Dementia with Lewy bodies, which are abnormal aggregates of protein that develop inside nerve cells, is thought to be second only to Alzheimer's disease as a cause of dementia. It is similar in some ways to both the dementia resulting from Alzheimer's disease and the movement problems of Parkinson's disease. While Alzheimer's disease usually begins quite gradually, dementia with Lewy bodies often has a rapid or acute onset. Typically there are recurrent visual hallucinations, and Parkinsonian motor symptoms such as rigidity and the loss of spontaneous movement. These patients will often have a sleep behaviour disorder that involves acting out dreams, including thrashing or kicking during sleep. Patients may also suffer from depression. As with all forms of dementia, it is more prevalent in people over the age of 65. It gets is name from the protein clumps that develop in nerve cells and damage them and it overlaps clinically with both Alzheimer's disease and Parkinson's disease, but is more associated with the latter. The overlap in the presenting symptoms – cognitive, emotional, and motor – can make an accurate differential diagnosis difficult.

Other forms of dementia may be caused by the reduced blood flow to the brain which occurs with ageing. Vascular dementia is a result of damage to the brain caused by problems with the arteries feeding the brain or heart. Symptoms begin suddenly, often after a stroke, and may occur in people with high blood pressure or previous strokes or heart attacks. As well as delivering oxygen and nutrients to the brain, the blood flow removes waste products in the fluid surrounding the brain, and these may include tau proteins and amyloid which are linked to Alzheimer's. Patients with dementia have evidence of reduced blood flow. A dramatic possibility to increase blood flow involves trepanation, making a hole in the skull which could alter the flow of fluids round the brain in a positive manner. It remains to be seen whether such a procedure can treat Alzheimer's, and whether it is acceptable.

Creutzfeldt-Jakob disease, which has been called mad cow disease, is another dementia. This is a rare and fatal brain disorder; most patients die within a year, and it usually occurs sporadically in people with no known risk factors. However, a few cases are hereditary or may caused by eating meat that has been infected. Signs and symptoms usually appear around the age of 60 and initially include problems with coordination, personality changes and impaired memory, judgement, thinking and vision. Mental impairment becomes severe as the illness progresses, and it often leads to blindness. Pneumonia and other infections are also common.

Mental illnesses with some similarities to Alzheimer's include Down's syndrome, which is due to an extra chromosome 21 in the sufferer's cells; patients have only two thirds of normal lifespan. HIV-associated dementia is not age-related, and results from infection with the human immunodeficiency virus, which causes AIDS, and leads to widespread destruction of brain matter which results in impaired memory, apathy, social withdrawal and difficulty concentrating. Often problems with movement also occur.

Parkinson's disease is the second most common neurodegenerative disorder usually occurring late in life and affects 120,000 people in the UK. It is due to the death of nerve cells that signal by the neurotransmitter dopamine, which activates cells in our brains that let us move, for reasons that remain unknown. It is characterised by debilitating symptoms of tremor, rigidity, and slowed ability to start and continue movements. Seventy-five per cent of all cases of Parkinson's disease begin after 60, and incidence increases each decade after that up to about 80 years of age.

Depression – which is characterised by negative thoughts, low self-esteem, lack of pleasure, and often physical symptoms – affects three times as many older people as dementia. It varies from mild to severe and affects 10–15 per cent of people over 65 living at home in the United Kingdom. However, the most common age for depression is around 45. There are twice as many depressed women as men. More than 2 million older people over the age of 65 in England have symptoms of depression, but the majority are not getting any help, according to a report by Age Concern. In the USA severe depression is present in 20 per cent of those over 85, and older people are, in fact, more likely to have mild depression than any other age group. This is not because older age is inherently depressing, but because depression is often a side effect of physical illness. It is the commonest and the most reversible mental health problem in old age.

The reasons for depression in old age may be different to those for younger age groups but usually involve a loss of some sort. Depression in old age can develop as a result of the complicated and hard events in life – the loss of relatives, loneliness, a change of lifestyle because of retirement, or the appearance of illnesses. Depression is the major cause of suicide. Four out of five suicides in older adults are men. Among men over 75, the suicide rate is around 15 per 100,000 and is similar to younger age groups. Depression can be treated with cognitive therapy and antidepressants. These treatments helped with my own severe depression, which occurred when I was 65. One of the causes for my depression was fear of retirement, but the main cause was anxiety about a heart problem.

Psychoanalysis is not helpful with respect to depression or dementias in old age, and this was even Freud's view. In 1905, showing a notably dismissive attitude to the old, he wrote: 'Psychotherapy is not possible near or above the age of 50, the elasticity of the mental processes, on which treatment depends, is as a rule lacking – old people are not educable – and, on the other hand, the mass of material to be dealt with would prolong the duration of the treatment indefinitely.'

Age alone does not cause sleep problems. Disturbed sleep, waking up tired every day, and other symptoms of insomnia are not a normal part of ageing, but pain and health issues are often obstacles to sleep for old people. A frequent need to go to the bathroom, arthritis, asthma, diabetes, osteoporosis, night time heartburn, menopause, and Alzheimer's can cause frequent awakenings.

Are there any mental gains that come with ageing? Wisdom can be one, along with the advantages of accumulated experience. Older adults are better at comprehension of questions, and detection of absurdities. They are able to give attention to quite complex tasks, including events requiring focused attention, and also when a task requires divided attention. But if things become very complex, they may do less well than the young. There is some evidence that discourse skills improve with age, and the elderly are capable of complex narratives. In spite of the declines mentioned earlier, older adults do very well performing their jobs. Knowledge about the job increases with age and is maintained. Many tasks become almost automatic. Computer skills are significantly less than those of the young, but brain scans have shown that using the internet boosts brain activity of the elderly more than reading, and this could help prevent dementia.

I talked recently with Dr Martin Blanchard, a geriatric psychiatrist, and asked how he got involved in old age psychiatry:

> I became interested in geriatric medicine when I was a student as it involved many disciplines, and I had a very good experience working in old-age psychiatry, the patients were so grateful. One of our problems in medicine is that we do not think enough about the quality of life, we prolong it. The main problem with our patients is not dementia but depression. There is no real treatment for dementia but rather there is management of the patients lives.

Did many of his patients actually want to die?

> That is in fact quite rare unless they have a severe depression. Even when frail and with problems they want to go on living. Few of our patients actually remain in hospital for more than several weeks. The number of patients we have to deal with has not increased over recent years, but the number of referrals we get from different GPs varies a great deal as they handle their patients in different ways.

Given the many problems, physical and mental, linked to old age, we need to look at how the old actually live.

4
Living

'Old age has its pleasures, which, though different, are not less than the pleasures of youth' – W. Somerset Maugham

The Greek poet Anacreon (*c.*572–488 BC) wrote one of the earliest poems about old age, and it strikes a cheerful note:

> Oft am I by the women told
> 'Poor Anacreon! thou growest old;
> Look; how thy hairs are falling all;
> Poor Anacreon, how they fall!'–
> And manage wisely the last stake.
> Whether I grow old or no,
> By the effects I do not know;
> But this I know, without being told,
> 'Tis time to live, if I grow old;
> 'Tis time short pleasures now to take,
> Of little life the best to make.

A happy old age is what many people spend their lives preparing for, particularly with regard to financial security and good health. But what is our lifestyle? How varied is it and is there much pleasure still to be had as one ages? Can one *enjoy* old age? This is an important question. There are at present 10 million in the UK over 65 and there will be double that in ten years' time. There are one million over 85.

Nobody wants to be old, but old age doesn't have to be a time of despair. Joan Bakewell offers a positive view: 'In their leisure time, the old aren't just boozing and cruising: the hardier spirits are climbing mountains, visiting the pole, meeting sponsored challenges. I have a friend in his late seventies who has recently taken up tap-dancing.' I interviewed Joan after she been asked to become 'the voice of older people':

> When I was 70 I wanted to reinvent myself, it was time to start something new. So I managed to start a column in the *Guardian* called 'Just Seventy'. It was up to me to have the idea – no one was going to come to me with it. My column was about being 70 and all the things you have to adjust to. For example, for women I wrote about them having to give up high heels, and children, and other changes. Also old women become socially invisible when they have lost their high heels. Sheila Hancock says she always asks for a corner table and then others around her will have been served before anyone has even brought her the menu. My columns were eventually put into a book which is still in print. And then the government in 2008 came to me – Harriet Harman phoned and said parliament was trying to outlaw ageism and would I be the voice of the older people. I only agreed to do it part-time as I wanted to continue to do my own work. I said I would pass on to her everything they tell me. I no longer do it.
>
> There are important differences between men and women as they get old. Their patterns diverge as men remain fertile, have children, have second wives, have a renewing life. Women know that they are no longer biologically needed and so they are in a psychological sense ready to grow old. I am rather against that. They can start to wear clothes designed for older people, sensible, rather neutral clothes that do not have any style to them, like your mother and grandmother did, but many are becoming more fashion conscious. I colour my hair as that keeps you looking a bit younger.

A lot of people worry about money, it's almost biological. They worry if they will have enough and where it is going to come from. The state pension is tiny and many have to live a lifestyle on a tiny amount of money. There is a sense of loss, things are not what they were – your children have flown the nest and your grandchildren have grown up. Living on your own as I do can present problems unless you have an attitude towards it. Many worry about how they will be right at the end of their lives and not being able to look after themselves – that is what I am making a TV programme about. Will they have to sell their house to pay for a carer, which is very expensive? Will they have to go into care? And that spoils their pleasure in being quite comfortable, having time on their hands, going out, playing golf. The absence of a competitive compulsion in life to do and achieve can make one much more relaxed. I do not get bored – too many books to read and films to see.

One feature of getting old is that your contemporaries die, and I have begun to make friends with younger people. New friendships are a blessing in old age. I am set on continuing to my mid-nineties – will keep working, travel – but I have signed the documents for non-resuscitation should I get very ill and go, for example into a coma. I am for euthanasia and support dignity in dying.

An important book that gives accounts of a diverse number of individuals' views of ageing is *About Time: Growing Old Disgracefully* by Irma Kurtz. She herself wrote: 'Talking to men and women of my generation, I am struck again and again about how we shed freight from that heavy goods vehicle, memory, as we age and gently drift back to early events that were the making of us. Growing old, as it separates us from the world, returns us to our original selves.'

Some researchers into the psychology and social aspects of ageing have distinguished between a third and a fourth age. In the third age, retirees from the work force are in relatively good health and are socially engaged, and it is a time of personal achievement and fulfilment – 'You're looking very well.' In the fourth age, usually over 85, there is the onset of most of the negative stereotypes of old age – functional breakdown of the psychological system, loss of positive wellbeing, psychological dependence on others, poor memory and impaired reasoning. Physical and mental deterioration are what we fear most, but in fact many over 85 are well and active, and many of today's pensioners enjoy a financial security unheard of in earlier generations. No association has been found between levels of mental ability when young and reported happiness when old. Quite the opposite has been found with health, as there is a high correlation between intelligence when young and good health when old.

Very old people rarely, it is said, covet status, rank or wealth. For many there is no longer the the problem of either looking for or having to work. There is much less anger and anxiety as one becomes more experienced, and understands so much more about life. Then there is the pleasure of becoming a grandparent, and the possibility of pursuing new interests Curiously, the old do not partake of the arts as much as those who are younger. Only a quarter of those aged over 75 have been to a museum or gallery in the last year. But equally, a quarter of those over 75 are involved in volunteering at least once a month for community activities.

Even at age 75-plus, a majority of people do not think of themselves as old, and many think of themselves as quite a few years younger. Perceptions do matter, and many are concerned that as they age they will lose respect and their health will deteriorate. Those who think of themselves as younger than their actual age have better health than those who think of themselves as older. Which comes first, the attitude to age or the better health, will only be settled as more longitudinal data become available.

ELSA (English Longitudinal Study of Ageing) found that about half the population of people 52 years and over describe ageing as a positive experience, and this contradicts a widely held belief that ageing is a negative process. But while ageing is described as negative by a minority, negative experiences of ageing are far more common amongst the poorest than the richest. Only one in five worry about growing older, but health is a key feature in their lives. The young perceive old age to start at 68, while the old see it as 75. Three fifths of those aged 80 and older were very positive about their health. A majority believe retirement is a time of leisure. The wealthier think it starts later than those not so wealthy. For many of the old 'ninety' is the new 'seventy' and these nonagenarians can be very active – travelling, learning and being with their family and friends. It is likely that 80-year-olds will in the future live as 60-year-olds live now.

Among Americans only 12 per cent said retirement would be the best years of their lives and about two thirds said their biggest concern about old age was becoming ill, and were afraid of losing their memory. About a quarter of those over 65 say they are in good or excellent health. Several surveys of the old show ageing is a positive experience for the majority: they do not think of themselves as old and feel younger than they are. Among those who would prefer it if they *were* younger, the mean desired age of those aged 65 was 42.

The elderly's level of religious participation in the US is greater than that of any other age group. For the elderly, the religious community is the largest source of social support outside of the family, and involvement in religious organisations is the most common type of voluntary social activity. Religious faith among older people effectively offers a sense of meaning, control and self-esteem, and helps in coping with the stress of old age. There is also some evidence to show that the religious live longer.

A Pew Research survey in the US asked about a wide range of potential benefits of old age. Good health, good friends and financial security predict happiness. Seven in ten respondents aged 65 and older said they were enjoying more time with their family; about two thirds enjoyed more time for hobbies, having financial security and not having to work. About six in ten say they get more respect and feel less stress than when they were younger. Daily prayer and meditation both increased with age. Among those aged 75 and older, just 35 per cent said they feel old.

The attitudes of the old to being old can, of course, vary widely. The interviews with elderly men in *Don't Call Me Grumpy* by Francis MacNab are revealing, and the advice in *Enjoy Old Age* by the psychologist B. F. Skinner and M. E. Vaughan is helpful. I find attractive this comment which they record as being given by several of the elderly: 'Thank God I no longer have to be nice to people.' There is also a noticeable tendency in some of the old to look back on their lives and try to make it coherent. And of course some say to themselves that life is no longer worth living: it is the same every day, what is the purpose of going on?

While the old experience declining health and the sense of being mortal, many maintain their wellbeing and are less troubled when exposed to negative emotions. Older adults are better at avoiding negative affect and maintaining positive affect, and they have better memory for positive pictures than negative ones. But for some old age is not good. I interviewed the Nobel Laureate novelist Doris Lessing, who is 91. What does she feel about being old? 'I feel irritated. I also do not feel as well as I should and I am not outside gardening. I am irritated that my health is not as good as it should be. I started feeing like this about a year ago. There are no good things about being old and I am short of everything. I am irritable and do not really like being irritable. My son is unwell as well. I would not like to go on living for long – it gets me down. I am not writing.' By contrast, the philosopher Mary Midgley, who is also 91 and who has just written another book, told me that when she gives a lecture she now enjoys the advantage of no longer minding what people think about her ideas.

The older pursue more emotionally meaningful goals, while the younger look for broader horizons. They are also less sensation-seeking. And they avoid physical risk, though can be fond of gambling. Many positively enjoy retirement and old age. In the arts there are many examples of creative people who have worked till they were old, albeit with some problems. Michelangelo, at 88, was designing the monumental dome of St Peter's Basilica; Stradivarius, in his 90s, produced two of his most famous violins; Verdi composed the opera *Falstaff* when an octogenarian. Bach and Beethoven were still creative composers in their old age. Rembrandt and da Vinci painted self-portraits that reflected their age; Goya at 70 made his look like a man of 50. Chateaubriand, the French writer and diplomat, so hated his ageing face he refused to have his portrait painted. But there was old Rembrandt with his penetrating self-portraits; old Titian's sensuous paintings of virgins; and Yeats's later works were his best. A recent pleasing example of the old being given key roles in a famous play was at the Old Vic in Bristol, where Sian Phillips at 76 played Juliet to Michael Byrne's 66-year-old Romeo. Judy Dench played Titania in a 2010 production of *A Midsummer Night's Dream* aged 75, for the 79-year-old director Peter Hall, 48 years after first stepping out in the role.

There has been a significant change in the sociology of ageing since the 1950s generation aged. They have a different perception, much influenced by greater affluence. There are more lifestyle choices, and those who were spenders when young continued as they aged. At the end of the century, retired people had more wealth than those of working age, and residential mobility had increased among the retirees. The common perception among advertising agencies is that younger age groups spend more than older age groups, but recent studies show consumers aged 65 to 74 outspend their counterparts in the 35-to-44 age group. About one in six women are now pensioners and this will probably increase to one in four over the next ten years. The idea that these women are becoming doddery and inactive is wrong. This is an image that relates to the situation 30 years ago. Now, to the contrary, the evidence is that many are active and young looking.

Older adults experience fewer stressful life events than

younger adults. They have, for example, less marital conflict and job stress. Old men are less critical of their bodies than women. A study or 340,000 Americans found that levels of stress began to decline in their early 20s and when they reached 50 then worry decreased, and happiness and enjoyment then increased till 85. It is almost like having, the researchers claimed, a new life that begins at 40. There is new wisdom and the old are better able to view their life circumstances positively.Those who age successfully are in good health, with high levels of mental and physical activity, and active involvement with their environment. Most older people take a holistic view of what 'health' means, including wellbeing and social factors, and in general take a positive view. Social contact remains a central issue for the aged, and there is a decline in interest in national issues, yet about two thirds of the aged turn out to vote at elections in the UK. More of them vote more than young people, and their vote is very important to all the political parties.

Adults are very capable of learning well into their 70s, which is a good reason to accept lifelong learning as more than just a pleasant mantra. Likewise, it seems beneficial for teachers in the higher educational setting to be aware of the differences between the older learner and the traditional college-age student. The differences are somewhat subtle, so it will take effort on the part of an instructor to understand and implement appropriate strategies. Learning in later life contributes to physical and mental health and wellbeing. It is also associated with increased self-confidence and community activity. But participation in further education for older people is very low, with only 10 per cent of the over 75s being involved. The focus is on the young, the under 25s; and just 1 per cent of the education budget is given to the old. There should be more funds to help those who are starting new careers as they age.

The University of the Third Age provides many opportunities for the elderly. It offers the chance to study over 300 different subjects in such fields as art, languages, music, history, life sciences, philosophy, computing, crafts, photography and walking, and the number increase each year. The membership of a typical University of the Third Age is about 250, but can be as small as 12 and as large as 2,000. Their approach is learning for pleasure, as there are no assessments or qualifications to be gained. Individual membership rose to over 230,000 in England in 2009.

Exclusion from computers and the web is particularly pronounced for older people, with only 30 per cent of people aged 65 and over ever using the internet. Computers are being modified for the old with larger power buttons and easy-to-read menus. But it has been suggested that computer games may be bad for the elderly as they can decrease participation in more effective lifestyle interventions such as exercise. Only 20 per cent of those aged 65 to 74 and just 7 per cent of over 75s do enough exercise – 30 minutes, five times a week.

When making choices about how to live, middle-aged and older adults attempt to preserve and maintain existing ways, and they prefer to accomplish this by using strategies tied to their past experiences. This may not always be wise, as a survey in 2006 of pensioners showed. They were asked what in their lives they would change if they could have their time again. While about one fifth would have married a different spouse, about one half would have saved more, and nearly three quarters would have had more sex. Old age can provide a useful excuse for men whose sexual abilities are failing. There are claims that the elderly get less pleasure from sexual intercourse, and they thus seek pleasure in erotic literature and the company of young women, and even voyeurism. Alison Park, co-director of the National Centre for Social Research's British Social Attitudes survey, says that on issues such as marriage, pre-marital sex and homosexuality, 'it doesn't follow that people become more restrictive in their attitudes as they get older. People's attitudes are shaped when they are quite young and stay with them.'

It is important to dispel the myth that as men get older their sexual abilities will significantly decrease. There's really no physiological or anatomical reason why a healthy man who takes good care of himself, and who doesn't have attendant medical problems, shouldn't be able to have a fulfilling and active sexual life. A comprehensive national survey of senior sexual attitudes, behaviours and problems in the United States has found that most people aged 57 to 85 think of sexuality as an important part of life and that the frequency of sexual activity, for those who are active, declines only slightly from the 50s to the early 70s, and that this activity continues into the 80s.

If the old have energy for sex, they also have sufficient energy for crime. In England and Wales, prisoners aged over 60 are the fastest growing age group in prison. The increase in the elderly prison population is due to harsher sentencing policies, which have resulted in the courts sending a larger proportion of criminals aged over 60 to prison to serve longer sentences. Between 1995 and 2000 the number of elderly males given custodial sentences increased by 55 per cent. In 2007 there were some two thousand prisoners aged over 60 in England and Wales, including about four hundred over 70. The majority of elderly men in prison were there for sex offences. The next highest offence was violence against the person, followed by drug offences. More than half of all elderly prisoners suffer from a mental disorder, mainly depression, which may be caused by or aggravated by imprisonment. In the USA, the number of prisoners over 50 is more than twice as many as a decade earlier.

One of the greatest pleasures of old age is having grandchildren – I have six. They usually need a minimum of care and they are a delight. Perhaps they will help look after me in my very old age – but I would not rely on it. Looking after grandchildren is a possible role for the old. Patsy Drysdale from Stranraer was crowned the UK's best grandparent in the 2008 Age Concern Grandparent of the Year Awards in association with Specsavers. The annual competition, now in its 19th year, is a celebration of how important grandparents are to family life. It gives grandchildren the chance to say thank you for all the love and support they receive from their grandparents. Patsy was shortlisted for the national award from hundreds of entries. She was nominated by her granddaughter Gina, aged 13, to thank her grandma for taking her in when it looked she was going to have to go into care following the death of her mother. Patsy has supported Gina through difficult times and Gina has been there when her grandmother needed her, nursing her after a cancerous tumour was removed from her lungs. In 2009 Christine Levin from Falmouth, Cornwall, was crowned the UK's best grandparent, again nominated by her granddaughter.

But there are also problems associated with the role of the caring grandparent. A study in London showed that children did better if they went to nursery school than if they were cared for by their grandparents. Their social skills at 3 years were worse and they had more behavioural problems already at 9 months, though their vocabulary was better. Worse still, it was found in the US that grandparents who looked after grandchildren or lived with them were in worse health. A recent study found that for many, friends and hobbies are more satisfying for the old than grandchildren.

With regard to the age of parents, researchers analysed the scores of children who had been tested at regular intervals in a variety of cognitive skills, including thinking and reasoning, memory, understanding, speaking and reading, as well as motor skills. Regardless of their mothers' ages, the older the fathers, the more likely the children were to have lower scores. By contrast, children with older mothers generally performed higher on the cognitive measures, a finding in line with most other studies, suggesting that these children may benefit from the more nurturing home environments associated with the generally higher income and education levels of older mothers.

There are great benefits to be had from pairing the elderly with pets, but there are also some risks and one has to be careful in the selection of a dog. Companion dogs can be very comforting and can bring much joy to any elderly person, and studies have proved that the overall wellbeing of old people can be improved when sharing love with a four-footed friend. Doctors, social workers, home care workers and nursing homes recommend companion animals to help the elderly, and this includes not just dogs but birds and cats. Dogs can provide more than just affection: studies have shown that they can lower blood pressure, offer a sense of security and safety, and decrease feelings of isolation. There is also good evidence that touch is very important to the wellbeing of humans. A cat curled up in the lap or the friendly touch of a dog's nose can give a sense of reassurance and satisfaction. Stroking a beloved pet can lower blood pressure and lift depression. The elderly will be kept active by feeding, grooming and caring for their pets. Dogs get them out of their living quarters and into the fresh air and sunshine, and this also helps them to get to know other people in the neighbourhood. Caring for a pet's needs gives the elderly an incentive to maintain their normal activities.

Ageing is more than an innate physical process; it also reflects patterns and choices made at individual and societal levels. The proportion of older people in England's rural areas is significantly higher than in urban areas. It is a trend that is likely to continue, as more people move to the countryside for quality-of-life reasons in their middle age, and stay on into retirement. Findings from Age Concern show that almost all older people in rural areas consider their local post office to be 'a lifeline', with over half of over-60s in the countryside fearing that post office closures would leave them more isolated. Rural post offices provide much more than just a postal service to older people. Many pensioners use their post office as a 'one-stop shop' to access their pension and benefits, pay their bills, get advice and information, and meet and socialise with others. Closures leave many older people increasingly financially and socially excluded.

A survey of nearly 14,000 people confirmed that the happiest older people are those living in the country. One in ten picked Devon as the best place for old people to retire to. Cornwall also got high ratings. Many were aspirational, seeing retirement as the start of a new life, as they may have 20 years left. Nearly half of over-50s plan to move when they retire and just 3 per cent thought of moving to London – less then 1 per cent thought it a desirable place to live, with the high cost of living being a significant factor. Many older people in London are afraid to go out and feel very isolated.

It is almost inevitable that with age there is an increasing likelihood of an individual living alone, and loneliness can be painful. About one in ten people aged 65 and over, the equivalent of more than one million older people, perceive themselves to be often or always lonely. Millions of elderly people do experience loneliness. Nearly half a million older people leave their houses only once a week and a further 300,000 are entirely housebound. Half a million of those over 65 spend Christmas day alone. Loss of local services such as post offices and small shops makes things worse. Children, partners and friends matter. Approximately twice as many people in the poorest wealth quintile compared with the richest feel isolated often or some of the time. Living alone, in turn, is more common in the poorer wealth groups. Not surprisingly, feeling left out is more common for people living without a spouse or with a spouse with whom they do not have a close relationship. Three in five women of 75 and older live alone, while less than a third of men of similar age do. Household size decreases with age more sharply for women than for men, with two thirds of women and one third of men aged 80 years and over living alone, compared to one in ten of both men and women in their early 50s. Many are home-owners, but many of the homes lived in are in a bad state. Over 80 per cent of older people want to stay in their own homes, which is hardly surprising, but about one half of those over 75 living in their own homes have a disability. Some 400 people aged 80 and over marry each year in the UK – more men than women, as the men marry younger women.

Those who are considered to be severely socially excluded belong to one or more of these categories: those aged 80 years and above, those who live alone, have no living children, have poor health, suffer depression, never use public transport, or do not own their accommodation. Social exclusion is also related to low income, those whose main source of income is via benefits, are unemployed, or take no physical exercise. Those who are socially excluded include some of the most deprived among the older population.

There is not much public effort to improve the lifestyle of the elderly in towns and with transport, but in London pensioners' playgrounds are planned for Hyde Park and other areas, with fitness equipment and an outdoor gym. These will be less intimidating and expensive than normal gyms. Buses need to be designed so the elderly can easily get on and off. A positive feature is that there are concessions for the fares of the old on public transport. Some 4 to 5 million in the US use mobility devices.

The lack of public lavatories makes it hard for the elderly in town centres, and unrepaired pavements can cause serious falls. In Japan, however, one fifth of whose population of 128 million is over 65, attempts are being made to cater for the elderly, including cars designed to be more responsive when the old are driving and even elderly porn. In the Sugamo region of Tokyo, the elderly flourish. Shop fronts have been modified to deal with wheelchairs and the goods in them are what the elderly need – including many medicines and walking aids. Most of the shops in Sugamo are barrier-free, giving easy access to people with canes, walkers or wheelchairs. Moreover, the layout of each shop is open and the height of the counter is rather low, and they provide an atmosphere where shop staff and customers can easily communicate. In the UK, a business network called Engage have established AGE OK to give credit to old-age-friendly products or schemes, the first being for remote controls for TV to help with poor sight.

There are currently about 1.5 million people of retirement age in full or part-time work, a significant increase in number. Those with middle incomes and wealth are the most likely to stay in work as they approach state pension age. The poor often stop work through ill-health or disability. Four out of five people with a compulsory retirement age in their job would not want to work beyond it. A survey in the US found that about half of those working beyond retirement age did so because they wanted to and only 17 per cent did so because they needed the money. Banks have been accused of deliberately misleading vulnerable elderly into gambling their savings in risky investments. It would be sensible for those over 70 to bring an adviser with them when thinking of such investments, and a senior manager should be involved.

Retirement only came to industrial societies in the

twentieth century, when people were living much longer – before that working lives mainly ended with death. A 65-year-old man can now expect to live another 16 years. As individuals approach retirement, they need to decide when to stop working and to examine their financial situation, particularly their pensions. Age for retirement varies, but 65 is common, and it affects the cost of pensions. State money for the old came after the Old Age Pensions Act in 1909, paying an amount of between 10p and 25p a week from age 70, on a means-tested basis. Then the Contributory Pensions Act in 1925 set up a state scheme for manual workers and others earning up to £250 a year – the pension was 50p a week from age 65. In 1946 the National Insurance Act introduced contributory state pensions for all. The basic state pension is a 'contribution based' benefit, and depends on an individual's National Insurance contributions, a system of insurance against illness and unemployment. For someone with the full number of qualifying years, typically 44 for a man and 39 for a woman, it is payable at a flat rate of £95.25 per week (2009/10). Less pension is paid if there are fewer qualifying years. The first report of the government's Pension Commission in 2004 outlined some of the main challenges facing UK pension provision; it suggested that either taxes will have to rise or people will have to work longer and save more, or face poverty in old age.

There is an old saying that old people yearn for retirement, but that many who have retired regret it. Ernest Hemingway said that retirement was the worst word in the English language as it indicated the loss of the activity that was at the centre of one's life. Denial of ageing can be very common. The restaurateur Antonio Carlucci sees retirement as death. Compulsory retirement below 65 is unlawful unless the employer can provide an objective reason. A worker can see their employment end at the age of 65 without any redundancy payment – even if they do not want to retire. However, there is a compelling case for the retirement age to rise, mainly so that the individuals can continue to earn money. The UK coalition government has recently decided to abolish the compulsory retirement age by October 2011. The young see the increase in the retirement age as blocking their own promotion, but not that there is a problem in how to financially support all those who have retired.

A survey suggested that about half of retirees found the current law satisfying, and only 7 per cent found it unsatisfying. One third said that spending more time with their families was a good reason for retirement, yet an increasing proportion of those in their mid-50s expect to be working after 65. Poor health plays a major role in deciding when to retire – more so than finance. But the nature of the job has an influence, as one third of those over 70 with jobs held managerial and professional positions.

The Employment Equality (Age) Regulations 2006 gave an employee approaching 65 the right to ask to continue working, but an employer can refuse without any explanation. The age regulations do not require an employer to give a reason for a refusal to grant an employee's request to continue working; the obligation is only to consider the request, to follow the correct procedures in relation to adhering to the time limits, and hold a meeting with the employee to discuss the matter. Some 25,000 are forced out of work each year for this reason. The Horndal effect shows how useful and competent older workers can be: production at a steel mill in Sweden went up 15 per cent as workers aged, with annual output per worker increased steadily for 15 years with no additional investment.

When the UK government scrapped the mandatory retirement age for civil servants from April 2010, they were lauded for being progressive. It would have been hypocrisy to send 65-year-old civil servants home for good, while the House of Commons was, before the recent general election packed with 89 MPs over the retirement age of 65. People over 60 are more active than ever before, and it is only right that the state recognise this. Many of the votes that keep MPs in office are from the over 65s. The House of Lords is often referred to as Britain's most expensive retirement home, since in 2010 the average age was 69. It can be argued that they have a significant collective wisdom that comes with age. Judges in the UK retire at 70, while in Canada, for federally appointed judges, retirement is mandatory at age 75, and in the USA Supreme Court judges have no retirement age and effectively have life tenure. An 89-year-old Supreme Court Justice recently commented: 'You can say I will retire within the next three years. I'm sure of that.' University professors in the US do not have a mandatory retirement age – lucky them. In Germany, a new law abolishing the compulsory retirement age of 68 for GPs and specialists in primary care recently came into effect.

Politicians can also work till they are old in other countries. But these days, not even in China do politicians work as long as they do in Italy. Former President Giorgio Napolitano was 84, and former Prime Minister Romano Prodi 70. In India – a young nation, where almost 75 per cent of their billion-plus population is below 40 years old, and over half have not even passed the age of 25 – some see it as ironical that most of their top politicians are in their 70s and some are over 80.

I have now, aged 80, found retirement quite hard. I miss my group of fellow scientists, mainly PhD students with whom I worked. I still have a room at University College and go in to seminars and very occasionally lecture. I am fortunately still invited to talk at various meetings, including some outside the UK. Most of my time is spent at home writing books, like this one. I do it lying on my bed with the computer on my lap. But I still pay tennis twice a week, jog slowly once a week, and cycle here and there. One of the pleasures of being a retired scientist is that I no longer have to apply for research grants and regularly publish good papers, or mark exam papers. But I do miss the research, even though I doubt that I am now competent to cope with the new technical advances in my subject, developmental biology. There are, for example, new techniques for identifying which genes are active in different places at different stages, which are now a bit beyond me. There are also, I regret, times when I wonder what the point of continuing to live really is.

Comparatively, Eastern civilisations have shown more respect to the old than those in the West. But even in India, where the old have not been seen as an eyesore struggling for existence as in some other societies, the elderly face a number of problems, such as poverty, illiteracy and inadequate health care. Most of the elderly in India are dependent on their children or close relatives. When the young leave home, there is a loss of sense of purpose to life. Youngsters dominate the workforce, with 20–35 being the desired age. Plans for 470,000 needy elderly to remain in their own homes will cost £670 million, and where will this money come from? And the definition of 'old' in other cultures can be very different: 40 may be considered over the hill, you do not stand a chance once you cross the landmark 50, and 60 is positively ancient! In China attitudes towards the elderly are more positive than those in the West, but a 2007 survey showed that student-age Chinese were less positive than the middle-aged.

In some societies the old are revered. In non-industrial societies the office of chieftainship is not infrequently occupied by aged persons, although in late life some of their authority and duties may be delegated to others. Among such people it is most unusual to reach the age of 65, so generally those of 50 are looked upon as being old. Important factors in societies where the old are respected include their active association with others, and assistance in their interests and enterprises. They can be regarded as repositories of knowledge, imparters of valuable information, and as having the ability to deal with the fearful supernatural powers. In societies without magic the attitude to the aged varies. The proportion of the old who remain active in these primitive societies is higher than in wealthier civilisations, for they utilise the services of their few old people. Probably nowhere has age received greater homage than among the Palaungs of North Burma, who attribute long life to virtue in a previous life. 'No one dares step upon their shadow lest harm befall him.' It is such a privilege and honour to be old among the Palaungs that as soon as a girl marries she is eager to appear older than her age. Examples of the glorification of old age in legends and stories are common in these societies.

Among the Zande in the Sudan magic predominates, and the old can have authority by virtue of their supernatural powers. Similarly among the Navajo in Arizona, magic gives the old authority. Memory can also give authority to the old, as among the Moslem Mendes in Sierra Leone; the chief must know the country's history and the lives and families of the founders. The Incas were a militant nation and everyone had to work from an early age; when over 50 they no longer had to do military service but continued to do useful work, sometimes to over 80.

But in most pre-industrial cultures – as Leo Simmons's book *Role of the Aged in Primitive Society*, on which this section is largely based, shows – life's last chapter has been a bitter one. There have been examples where the old were actually killed off. Surviving folklore reflects widespread resignation as to the inevitability of impoverishment, failing health and vitality, and the loss of family and community status. Such euphemisms as 'golden years' and 'senior citizens' rarely exist. Many primitive societies did not encourage the survival of the old; either they were left to their lot or sacrificed. Among the Yakuts, who live a semi-nomadic life in Siberia, life was very hard, and the father dominated the family until old age made him feeble, at which time the sons took over and treated him almost as a slave.

The extreme authority of aged fathers over their descendants is not uncommon. Nevertheless, there are cases where very aged parents are pitifully abused by their children and other relatives. In general, both old men and old women tend to receive better care in agricultural societies, where residence is permanent and where the food supply can be kept more or less constant. Among the Hopi in Arizona, a herding and farming people, old men tend their flocks until feeble and nearly blind. When unable to go to the fields any longer, they sit in the house where they can do handiwork like weaving blankets, or making sandals. The old frequently express the desire to 'keep on working' until they die. Among the Hopi there are many accounts of the amazing powers and exploits of old people.

One observer notes: 'Retirement is impossible at any age.' The old are less useful in societies characterised by collecting, hunting and perhaps fishing, as these are not skills they any longer have. For the old without children or wealth everywhere, times can be hard.

5
Curing

'All diseases run into one, old age' – Ralph Waldo Emerson

The hieroglyphic for 'old' in ancient Egypt in 2800 BC was a bent person leaning on a staff – perhaps the first depiction of the ravages of osteoporosis. Like all humans, our ancestors wanted to know the cause of the things that affected them, and so it was with ageing. They were also looking for ways to prevent it. Different from the legends about immortality, which will be discussed later, were the theories as to why ageing occurred and how to avoid it, and these go back a long time. Originally it was the function of religious beliefs to provide such explanations. Eventually such enquiries led to the true study of ageing and how it might be treated – geriatrics.

As long ago as 1550 BC the Ebers papyrus from Egypt, one of the oldest preserved medical documents, suggested that debility through senile decay is due to accumulation of pus in the heart. This is probably the first non-religious explanation of ageing. The Taoists in ancient China believed that ageing was due to the loss of some vital principle which they equated with the loss of semen in men, and thus taught secret techniques by which men could have an orgasm without ejaculation. Such men, they claimed, would age much less. Plus, if you learned to undertake effortless action, take vital breaths and eat magical foods such as ginseng, you could also slow down the ageing process. In Ayurvedic medicine in India the ancient sage Maharishi Chyavana propounded his idea of anti-ageing therapy. When the Maharishi was bogged down by old age and low energy levels he started taking chyavanaprasha, an astonishing tonic and anti-ageing medicine, and claimed he soon found himself on the road to complete recovery. Gooseberry is the main constituent of this tonic.

Around 400 BC, medicine in Greece was steeped in religious belief. Illness and ageing would automatically be attributed to the activities of gods or demons. But some of the earliest non-mystical explanations for ageing also came from the Greeks. The 'father' of medicine, Hippocrates, instead of ascribing diseases to divine origins, discussed their physical causes. He believed that certain diseases afflicted certain ages. His theory was that ageing was due to loss of heat and moisture. Aristotle, who had a very negative view of the old, perceived the aged body as dry and cold, and also thought that old age was due to the diminishing of heat in the body, as heat was the essence of life generated by the heart. Galen, for whom old age was due to the dry and cold constitution of the body, recommended that the old should take hot baths, drink wine and be active. For Galen old age is not a disease, and is not contrary to nature. For St Augustine, some 800 years later, illness and ageing were the result of Adam and Eve being driven out of the Garden of Eden.

It must be remembered that there was no chance in those times of acquiring any scientific understanding of ageing, as it was only some two thousand years later, in the nineteenth century, that it was at last discovered that the body was made of cells. Darwin's theory of evolution was also of great importance, as we shall see.

The Arabic philosopher Avicenna (981–1037) followed Galen and saw no way to prevent the drying out that caused old age. *The Canon of Medicine*, written by Avicenna in 1025, was the first book to offer instruction for the care of the aged, foreshadowing modern gerontology and geriatrics. In a chapter entitled 'Regimen of Old Age', Avicenna was concerned with how 'old folk need plenty of sleep', how their bodies should be anointed with oil, and recommended exercises such as walking or horse-riding. One thesis of the *Canon* discussed the diet suitable for old people, and dedicated several sections to elderly patients who become constipated. The Arabic physician Ibn Al-Jazzar (*c.*898–980), also wrote a special book on the medicine and health of the elderly.

Roger Bacon (*c.*1214–94), a Franciscan friar, was the first to propose a scientific programme of epidemiological investigations into the longevity of people living in different places, and under different conditions. He also noted that the pursuit of knowledge depended on 'the fresh examination of particulars', and that there needed to be a systematic observation of nature. He wrote a book on ageing in which he suggested that old age could be warded off by eating a controlled diet, proper rest, exercise, moderation in lifestyle and good hygiene. So far so good, but he also suggested inhaling the breath of a young virgin. Following a common theme in those times, that ageing was the result of the loss of some vital material, Bacon claimed that the breath of young virgins could replenish the loss of this vital essence. This belief most probably came from the biblical story of King David sleeping between two virgins when he was old to restore his youth, though not necessarily having sex. A young virgin could preserve a man's youth because the heat and moisture of the young woman would transfer to the old man and revitalise him.

Roger Bacon also claimed that life could be extended and that Methusaleh was an example, that the neglect of hygiene shortened life, and some individuals had used secret arts to prolong their lives. One example was a farmer who drank a golden drink he found in the field and lived a long time, and this supported Bacon's alchemical convictions. In Italy, in the mid-sixteenth century, Alvise Cornaro said that life could be extended by eating less as it used up less innate moisture, and it was necessary to keep the four humours that had been the basis of Greek medicine – blood, phlegm, yellow bile and black bile – nicely balanced. There was still no serious science of ageing.

It was Francis Bacon, a key promoter of the renaissance of science and author of *The History of Life and Death* (1638), who first proposed a study of ageing in order to find out its causes and how to prevent it. He was the first to acknowledge the prolongation of life as an aim of medicine. He argued that ageing was a complex process, yet capable of remediation, but 'It is natural to die as to be born.' He did not believe that old age was due to a loss of some vital substance. He made little progress, but he did recommended exercise. He also apparently had a light touch: 'I will never be an old man. To me, old age is always 15 years older than I am.' There is a story that while travelling in a coach towards Highgate in London he concluded that cold might prevent ageing. He tried the experiment at once, stopped the coach, bought a hen, and stuffed its body with snow. But the cold affected him and he died a few days later. His interest in ageing was an important stimulus to its being studied by others.

Dr George Cheyne, an eighteenth-century doctor, believed the English were dying due to an excess of comfort, wealth, and luxury – the 'English Malady' – and that the way to prevent ageing was by eating only enough food to allow the body to maintain its heat. A little later the German physician Hufeland argued that fast living led to short living, that you should drink no alcohol, chew your food deliberately, and be positive. He stated: 'We frequently find a very advanced old age amongst men who from youth upwards have lived, for the most part, upon the vegetable diet, and, perhaps, have never tasted flesh.' He accepted the view that at birth an individual was endowed with a finite amount of vitality and that this decreased with age.

The scientific study of ageing only began to make progress with the work of Benjamin Gompertz, whose paper in 1825 described human vital statistics from several countries, and showed that the prevalence of many diseases increased in the same way as mortality. He concluded that death may be the consequence of two generally coexisting causes; the one, chance, without previous disposition to death or deterioration; the other, a deterioration, or an increased inability to withstand destruction, namely ageing. Gompertz was interested in the latter situation: how can we model the probability of a person living to a certain age, if nothing unexpected happens to him? His important results showed that mortality increases exponentially as age increases between sexual maturity and old age.

The Belgian scientist Adolphe Quetelet recognised that both social and biological factors determined how long humans live, and made important contributions to life histories. Quetelet began his research by the physical study of the 'average man'. He laboriously recorded population statistics surrounding the birth, height and physical proportions of men at various ages. Among his findings were strong relationships between age and crime. Charles Darwin's theory of evolution came out at this period, and Quetelet wanted to know whether selection for those who are better adapted continued after the individual's reproductive phase. Darwin's cousin, Francis Galton, was interested in ageing, and collaborated with Quetelet to measure the correlation between age and strength. In 1884 he collected the physical reaction times of some 9,000 people aged from 5 to 80, which were not analysed until much later.

In 1881 August Weismann delivered an important lecture on ageing at the University of Freiburg. It was the first attempt to explain ageing in terms of Darwinian evolution and the behaviour of cells. He was convinced that immortality would be a useless luxury and of no value to an organism, and that the cause of ageing would be a limitation of cells' ability to reproduce. He regarded ageing as adaptive, as it helped get rid of decrepit old individuals who competed for resources with others in their group. This was wrong, as we shall see. He nevertheless recognised the important principle that once an individual had successfully reproduced and cared for their offspring, it ceased to be of any value to the species. He also made clear that the germ-line cells which give rise to eggs and sperm must not be subject to ageing, for if they were, the species would die out. It was another 60 or so years before there were further attempts to understand the evolution of ageing.

Jean-Martin Charcot, a famous neurologist at the Pasteur Institute, also promoted the study of old age, which he recognised as being neglected. His lectures on the medicine of old age, *Clinical Lectures On Senile And Chronic Diseases*, aroused scientific interest in the field, and became available in English translation in 1881. They had a big influence, as Charcot saw old age as the simultaneous enfeebling of function and a special set of degenerative diseases, and these needed to be distinguished. Elie Metchnikoff, a Russian who went to the Pasteur in Paris in 1888, continued Charcot's work and coined the term gerontology in 1903. *Geronte* is French for 'man' and has nothing to do with ageing but it remains with us as a name for the science of ageing.

Metchnikoff won a Nobel prize for showing how certain cells in our bodies defend us against invaders like bacteria, by eating them and dead material, a process known as phagocytosis, and he saw old age as cellular involution in which cell decay outbalances cell growth. He believed ageing was due to bacterial toxins released from the intestine, and that Bulgarians lived especially long lives because they ate yogurt. He thus touted yogurt as an anti-ageing medicine. Based on his theory, he drank sour milk every day. George Edward Day (1815–72) wrote a common-sense book from the physician's perspective on ageing in 1848. He complained that other physicians had little interest in caring for the ills of the aged. That refrain still rang true during the first few years of the twentieth century.

Modern geriatrics was born with the invention of the word 'geriatrics' by Ignatz Leo Nascher from the Greek word *geras* for age. Nascher was born in Vienna in 1863, graduated as a pharmacist and then obtained his medical degree from New York University. He wrote a number of articles on geriatrics and a 400-page book, published in 1914, *Geriatrics: The Diseases of Old Age and Their Treatment*. He described ageing as a process of cell and tissue degeneration. He thought, mistakenly, that all our cells except for the brain were replaced as we aged. A major problem for him was how to distinguish between diseases *in* old age and diseases *of* old age. His interests in geriatrics and his development of treatments for older people almost certainly came from visits to Austria, where the care of elderly people was blossoming at the time. He retired at the age of 66.

Nascher's interest in geriatrics is a bit strange as it contrasted wildly with his contemporary William Osler, the famous Canadian physician who was chairman of medicine at Johns Hopkins in Baltimore. Osler appeared to be remarkably ageist, as shown in his final address, called 'The Fixed Period', in which he stated that men over 40 years, beyond the golden age of 25 to 40, were relatively useless. Men over 60 years were considered absolutely useless, and chloroform was not a bad idea for this age group. This address is said to have been responsible for a number of suicides.

While there were scientific studies on child development, ageing was still largely ignored in the early twentieth century. The psychologist G. Stanley Hall was a founding father of psychology as a science. His major work was on child development, but, concerned about his own ageing, he wrote a book about ageing, *Senescence*, in 1922. He interviewed some elderly adults and found that their attitudes towards death changed as they aged. This was the first analysis of the changing attitudes and thinking linked to ageing:

> How different we find old age from what we had expected or observed it to be; how little there is in common between what we feel toward it and the way we find it regarded by our juniors; and how hard it is to conform to their expectations of us! They think we have glided into a peaceful harbor and have only to cast anchor and be at rest.

It was Peter Medawar in 1952 who pointed out that environmental factors progressively reduce an individual's lifespan, and natural selection would ensure that the good genes that support reproduction act early, and the bad ones that prevent reproduction much later. This was a major advance and it later became the basis for Tom Kirkwood's disposable soma theory, which recognised that just a small amount of energy was devoted to repair of ageing processes as compared to reproduction, growth and defence. The theory also claims that ageing is due to the accumulation of damage to the body, and that long-living organisms devote more to repair.

Perhaps the greatest impetus for the modern 'merchants of immortality' came from Leonard Hayflick's finding that there were just a finite number of times a fibroblast cell could divide when placed in culture. This eventually became known as the 'Hayflick Limit'. The original article by Hayflick was rejected by the *Journal of Experimental Medicine* with a scathing letter from the editor that stated, in part, 'The largest fact to have come from tissue culture research in the last fifty years is that cells inherently capable of multiplying, will do so indefinitely if supplied with the right milieu in vitro.' It was eventually published in *Experimental Cell Research* in 1961.

If Nascher was the father of geriatrics, Marjory Warren was its mother – particularly in relation to care of the aged. She worked at the Isleworth Infirmary, which in 1935 took over responsibility for an adjacent workhouse to form the West Middlesex County Hospital. During 1936 Dr Warren systematically reviewed the several hundred inmates of the old workhouse wards. Many of the patients were old and infirm, and she matched care to their needs. She initiated an upgrading of the wards, thereby improving the morale of both patients and staff. She advocated creating a medical speciality of geriatrics, providing special geriatric units in general hospitals, and teaching medical students about the care of elderly people. Among her innovations was to enhance the environment and emphasise increased motivation on the part of the patient.

Before the Second World War there had been little interest in old peoples' mental or physical health. Joseph Sheldon, while working at the Royal Hospital in Wolverhampton, undertook a survey of 583 old people sponsored by the Nuffield foundation, which he published in his book *The Social Medicine of Ageing* in 1948. He found that over 90 per cent were living at home and many had severe problems with respect to care. He introduced home physiotherapy and promoted environmental modifications to prevent falls, which were all too common. Old-age psychiatry was only recognised as a speciality by the Department of Health in 1989. Now, of the 1700 patients each GP typically has, about 6 per cent are over 75 and 2 per cent over 80. There will be around six consultations a year for the over-65s, so old age is quite a burden for GPs.

The first chair for geriatrics in the world was the Cargill Chair at Glasgow University, awarded to Dr Ferguson Anderson in 1965. Alex Comfort, more famous perhaps as a novelist and for writing *The Joy of Sex*, was a great propagandist for research on ageing. His early research was on ageing in the fruit fly *Drosophila* and thoroughbred horses. He then attempted to determine biomeasures of physiological ageing.

In the US the first head of the Unit on Aging within the Division of Chemotherapy at the National Institutes of Health, Nathan Wetherwell Shock, was appointed in 1940. In 1948, the gerontology branch was moved to be under the National Heart Institute. An attempt was made to have an Institute of Ageing established with Heart as a subsidiary, but this failed, as a physician to the Senate stated, 'We don't need research on Ageing. All we need to do is go into the library and read what has been published.' This contrasts with Nathan Wetherwell Shock's own viewpoint, enunciated just before his death in 1989: 'I would remind you that we were formed and nurtured in the firm belief that the biological phenomenon we call "ageing" was worthy of scientific pursuit. We have achieved some degree of success. I would caution, however, that our future will be determined only, and only, by the quality of our scientific research on understanding the basic mechanisms of ageing processes.' In 1974, Congress granted authority to form the National Institute of Aging to provide leadership in ageing research and training.

Research on ageing expanded significantly as it was realised that life expectancy was increasing, and thus the number of elderly. The Gerontological Society of America was founded in 1946 and the field has grown very fast. There are now many scientific journals devoted to the topic, such as *Gerontology* and *Age and Ageing*. But compared with certain fields of medical research, this topic is still relatively neglected. In the words of Professor Tom Kirkwood:

> I think doctors struggle with ageing as their training is to diagnose and treat diseases – they want to cure someone. For them ageing is a medical failure. There is a little progress with age-related diseases. It may be possible to cure Alzheimer's but this is very difficult and prevention is more promising. There has not been enough research on very old people, which is what we are doing in Newcastle. Not a single person in our study over 85 has zero age-related disease, most have four or five.

6
Evolving

'Getting older is no problem. You just have to live long enough'
– Groucho Marx

Research into the nature of ageing has helped us to understand its mechanisms first in terms of evolution, and then cell behaviour. We are essentially a society of cells and all our functions are determined by the activities of our cells. Evolution plays a key role, since it has selected cells to behave in a way that gives organisms reproductive success – a fundamental feature of Darwinian evolution. Evolution is not interested in health but only in reproductive success. Almost all the features of an organism, including of course humans, have been selected on this basis. The fertilised egg gives rise by division to all the cells that make up our body, as well as those of all other animals. Genes are turned on and off during the development of the embryo and this determines when and where particular proteins are made in cells and so also their behaviour. The details of this process have been selected for during evolution to give rise to adults that will reproduce.

Has ageing also been selected, and is it adaptive in that it helps reproduction? There have been suggestions that ageing was selected in order to reduce the number of adults so they did not compete with each other and so reduce reproduction in the group, but there is evidence to show that this is wrong.

It is essential to distinguish changes with time as an organism develops, and later grows, from the process of ageing. Ageing is not similar to the other biological changes that we go through with time as we develop in the embryo, and then grow older and mature after birth. An embryo gets older from the time it is fertilised, and the most obvious change with age after birth is growth itself, which is part of our genetically controlled developmental programme. We continue to grow for some 16 years. Puberty begins around 11 years and is the period of transition from childhood to adolescence, marked by the development of secondary sexual characteristics, accelerated growth, behavioural changes, and eventual attainment of reproductive capacity. Puberty changes occur as a consequence of the activation of a complex system that leads to an increase in frequency and amplitude of the hormones which stimulate the growth of sexual organs. This system is active in the early infancy periods, but becomes relatively quiescent during childhood, and puberty is marked by its reactivation leading to sexual maturity.

A remarkable case of failure to grow is Brooke Greenberg, a girl from Maryland who at 17 years old remained physically and cognitively similar to a toddler, despite her increasing age. She was about 30 inches tall, weighed about 16 pounds and had an estimated mental age of 9 months to 1 year. Brooke's doctors termed her condition Syndrome X.

Another major change with age is that each of us will have two successive sets of teeth. The baby teeth begin to erupt at the age of around six months. Usually by 2 years old most of a child's baby teeth will be in place. Some children get their teeth early, others later. Then typically by the age of 12, all of a child's baby teeth will have fallen out and been sequentially replaced by a second set of teeth.

All these changes with age are quite different from ageing with its negative effects, and have been selected in evolution as part of our development programme to help with reproduction. So why do we have the negative effects of ageing? Was ageing selected and programmed into our development?

The blame must fall heavily on evolution. To repeat, evolution is only interested in reproduction and not in health once we have reproduced. Ageing, as we shall see, is due to the accumulation of damage in our cells with time. Ageing is not part of our developmental programme and there are no normal genes that promote ageing, though as we shall see there are changes in genes which can cause premature ageing. On the contrary, evolution has sensibly selected cell activities that prevent the damage in cells due to ageing, but which are usually only active until reproduction is greatly reduced. No animals die of old age, but they die because of predators and illnesses, including those which are age-related. The effect of evolution can be seen by comparing two-year-old mice with baby elephants at the same age. The mice are already old. Evolution has selected mechanisms to prevent the elephant ageing before it has offspring, and for some elephants old age is only evident from worn-out tusks. Evolution has generated great diversity in lifespan. For example, rats live for three years, squirrels 25.

As mentioned earlier, August Weismann, the great

German theorist and experimental biologist of the nineteenth century, was one of the first biologists to use evolutionary arguments to explain ageing. His initial idea was that there exists a specific death-mechanism designed by natural selection to eliminate the old, and therefore worn-out, members of a population. The purpose of this programmed death of the old is to clean up the living space and to free up resources for younger generations: '. . . there is no reason to expect life to be prolonged beyond the reproductive period; so the end of this period is usually more or less coincident with death.' Weismann probably came to this idea while reading the following notes of one of Darwin's contemporaries and a co-discoverer of natural selection, Alfred Russel Wallace, which he later cited in his essay 'The Duration of Life':

> . . . when one or more individuals have provided a sufficient number of successors they themselves, as consumers of nourishment in a constantly increasing degree, are an injury to those successors. Natural selection therefore weeds them out, and in many cases favours such races as die almost immediately after they have left successors.

But the theory is wrong, as almost all animals in the wild die before they get old. Death in the natural environment is not caused by ageing but is due to many other factors, particularly predators. Some animals like elephants do age in the wild, but such cases are rare. Wild mice die in the field at about 10 months, while in the laboratory they can live for several years. A number of animals have lifespans longer than might have been expected – for example flying birds live three times longer than land-living animals. Robins live 17 years but the albatross 50 years.

This is because flying enabled them to escape predators and find new food sites, so early reproduction was no longer necessary. Why some reptiles like crocodiles and turtles have long lives is not clear.

The illnesses associated with ageing have a significant negative impact on human mortality. Weissman later rejected his theory, and then wisely proposed that ageing was the result of resources being given to the germ line rather than the body. If deleterious ageing occurred in germ cells, eggs or sperm, the species would die out – how right he was.

Theories concerning the ageing process emerged which are not based on it being adaptive, and thus not due to pressures of natural selection. The first was the 'mutation accumulation' theory, first proposed by the great scientist Peter Medawar in 1952, and referred to earlier, which proposes that mutations in the DNA of genes which lead to detrimental age-related changes in cells could accumulate over successive generations, if their serious negative effects were only expressed well after the age of peak reproductive success. These mutations are chance events. Life tables for humans show that the lowest likelihood of death in human females comes at about the age of 14, which in primitive societies would likely be an age of peak reproduction. Evolution has ensured that the peak of reproduction is when animals are young. Women lose their eggs at a more or less constant rate until they are 35, when the rate increases twofold.

Deleterious mutations expressed later in life are relatively neutral to selection because their bearers have already reproduced, and so have transmitted their genes to the next generation. As few individuals would actually reach those ages, such mutations would escape negative selective pressure – evolution would neglect them. The theory also predicts that if there are fewer external hazards for an animal, ageing will be slowed down, as is the case for animals like the albatross. According to this theory, ageing is a non-adaptive trait because natural selection is negligent of events that occur in a few long-lived animals that provide little additional contribution to offspring numbers.

Genes can be beneficial in early life, and then damaging later on. In other words, genes showing favourable effects on fitness at young ages, and deleterious ones at old age, could explain the ageing process. Such genes will be maintained in the population due to their positive effect on reproduction at young ages despite their negative effects at older post-reproductive ages, and those effects in later life will look exactly like the ageing process.

Mutation accumulation theory thus suggests that from an evolutionary perspective, ageing is an inevitable result of the declining force of natural selection with age. For example, a mutant gene that kills young children will be strongly selected against and so will not be passed to the next generation, while a lethal mutation with effects confined to people over the age of 80 will experience no selection because it has no effect on reproduction, and people with this mutation will have already passed it to their offspring by that age. Over successive generations, late-acting deleterious mutations will accumulate, leading to an increase in mortality rates late in life, which is just what we see and experience.

According to this theory, persons loaded with a deleterious mutation have fewer chances to reproduce if the deleterious effect of this mutation is expressed earlier in life. For example, patients with progeria, a genetic disease with symptoms of premature ageing, live for only about 12 years and, therefore, cannot pass their mutant genes to subsequent generations. In such conditions, the progeria is only due to new mutations and is not from the genes of parents. By contrast, people expressing a mutation at older ages can reproduce before the illness occurs; such as is the case with familial Alzheimer's disease. As an outcome, progeria is less frequent than late diseases such as Alzheimer's because the mutant genes responsible for the Alzheimer's disease are not removed from the gene pool as readily as progeria genes, and can thus accumulate in successive generations. In other words, the mutation accumulation theory correctly predicts that the frequency of genetic diseases should increase at older ages.

A second theory postulates that there might be genes whose expression is harmful in later life, but which are not silent earlier in life because they are actually beneficial to survival or reproductive fitness, and have some beneficial effects. Such mutations could thus have a selective advantage in early life and then a negative one later on. These genes will be maintained in the population due to their positive effect on reproduction at young ages despite their negative effects at old post-reproductive age, and their negative effects in later life will look exactly like the ageing process. Suppose, for example, that there is a gene increasing the fixation of calcium in bones. Such a gene may have positive effects early in life because the risk of bone fracture and subsequent death is decreased, but such a gene may have negative effects later in life because of increased risk of osteoarthritis due to excessive calcification. In the wild, such a gene has no actual negative effect because most animals die long before its negative effects can be observed. There is thus a trade-off between an actual positive effect at a young age, and a potential negative one at old age; this negative effect may become effective only if animals live in protected environments such as zoos or laboratories. Costly ornaments of male birds to attract females are essential for reproduction but a burden in later life – peacocks have limited mobility.

Although these concepts as to how mutations can cause ageing have guided attempts to merge evolutionary theory with empirical studies of the biology of ageing, there is little evidence of cumulative mutations that give rise to ageing, and only rare examples of genes that display the necessary early and late functions have been found. These theories do explain the universal occurrence of ageing. But they do not explain the actual process of ageing.

Ageing is best understood as the result of accumulation of random molecular damage in cells for a variety of causes – essentially errors due to wear and tear, and the mechanisms that cause this, and that involve damage to genes and proteins which the cells are unable to reliably repair, will be discussed next. These chance events occur in all body cells, and there are some mechanisms to repair the damage. An exception to such damage is in the germ cells that give rise to the next generation. Germ cells dare not suffer age-related damage, as if they did there would soon be no future healthy offspring. Evolution knows this and ensures that they do not age. By contrast, body cells do age, and evolution only cares to limit this so that reproduction can occur. Evolution selects those cellular activities that delay ageing until reproduction is completed.

Explaining ageing in these terms is partly based on an idea of Weismann, who dropped his theory that ageing was adaptive, and then suggested that ageing evolved because organisms separate in their body those organs involved in reproduction, particularly those that give rise to germ cells – eggs and sperm – from the rest of the body. They invest heavily in those organs involved in reproduction, and this neglect of the body results in ageing. Support for this is found in model organisms, where fertility and lifespan are closely linked. In the nematode *C. elegans*, cutting out of germline precursor cells of the gonad abolishes reproduction but extends lifespan, as do mutations that reduce germline proliferation. In the fruit-fly *D. melanogaster*, a reduction in reproduction extends lifespan in females, and certain long-lived mutant females exhibit reduced egg laying, with some being almost sterile. Certain mice that have mutations causing dwarfism are long-lived and sterile.

Researchers have also found that ageing and lifespan do evolve in subsequent generations of biological species in a theoretically predicted direction, depending on particular living conditions. For example, selection for later reproduction – artificial selection of late-born progeny for further breeding – produced, as expected, longer-lived fruit flies, while placing animals in a more dangerous environment with high extrinsic mortality redirected evolution, as predicted, to a shorter lifespan in subsequent generations. Selection of eggs from older flies progressively led to much older flies which lived twice as long.

This all fits with Thomas Kirkwood's disposable soma

theory, where soma refers to the body. The power of selection fades with age. The disposable soma theory argued that 'it may be selectively advantageous for higher organisms to adopt an energy saving strategy of reduced accuracy in somatic cells to accelerate development and reproduction, but the consequence will be eventual deterioration and death'. Given finite resources, the more the body spends on maintenance of the body, the less it can spend on reproduction. Molecular proofreading is reduced and so are other accuracy-promoting devices in body cells. Energy must be devoted to germ cell reliability but damage can accumulate in body cells – there are so few germ cells by comparison. From the point of view of evolution, the prevention of ageing is only necessary until the animals have reproduced and cared for the young sufficiently well; nature has therefore provided repair measures to delay the process until that is done. According to this theory, we and other animals are disposable once reproduction and the rearing of children have been completed.

Pacific salmon of both sexes do not care for the young and they die a few weeks after spawning. The male marsupial mouse dies after intense spawning from immune system collapse, but not the female. There are also animals that live well past their reproductive period – including whales and human females. In both cases this is due to their looking after and nursing the young, their own as well as those of others in the case of whales.

There is overwhelming evidence that there are strong genetic influences on the rate of ageing. Perhaps the most compelling evidence is that the differences of rates of ageing within individuals of a species are negligible compared with the vast differences across species. Honeybee workers live only a few weeks compared to the queen, who lives for years because she was fed honey when a larva. A mayfly moults, reproduces and dies within a single day, in some cases with a functional lifespan measured in hours; by contrast, giant tortoises can live for over 150 years, helped probably by their protective armour. The powerful influence of genetics is further reflected by the ever increasing number of single-gene mutations that can influence the lifespan of organisms ranging from yeast to mice.

An important example is that of female reproduction changing with getting older. This, due to menopause, is unlike ageing, and is programmed by our genes. Women can reproduce over long periods. The oldest mother is from India – she had twins at 70 with IVF. In the UK the oldest is 66. It is argued that 63 should be the maximum age, as the child needs a mother for some 20 years, which takes her to 83. A girl became the UK's youngest mother at the age of 12.

This raises the question of why there is a menopause in women and thus an end to reproduction. The average age in Britain for the menopause to occur is 51 years old. Why do women forgo years of their reproductive lives? What selection pressures could result in this unique human adaptation? Menopause may be explained by the 'good mother' theory – energy should be devoted to looking after children rather than having more. It may have been that, since childbirth is risky in humans, menopause allowed older women to survive longer and better raise their existing children. Another possibility is commonly known as the 'grandmother' hypothesis, and argues that women who stopped ovulating in their golden years were freed from the costs of reproduction and were better able to invest in their existing children and grandchildren, thus helping to ensure that more individuals with their menopause-inducing genes thrived and had children themselves.

A remarkably complete and instructive data set from Gambia offers a window into a world without the benefits of modern health care. What the data reveals is that children were significantly more likely to survive to adulthood if they had a grandmother's assistance. Grandmothers from Gambia are crucial to infant survival. In other studies data revealed that a child was over 10 times less likely to survive if its mother died before it was two years old, but that children between one and two had twice the chance of surviving if their maternal grandmother was still alive. No other relatives had any effect. But while menopause may result in less cancer, it increases the risk of heart disease and osteoporosis.

7
Understanding

'From age to age, nothing changes, and yet everything is completely different' – Aldous Huxley

If ageing is not programmed by our genes, then why and how do we age? The answer lies in our cells. We are essentially a society of billions of cells. Cells, for their size, are the most complex structures in the universe. It is proteins that determine how cells behave; genes only provide the essential code for making proteins. A typical cell, like one in our skin, will contain thousands of different proteins, and millions of copies of some of them. Complex interactions between the proteins and the genes determine which proteins will be synthesised and so determine how the cell behaves.

Proteins are long strings of quite small units, amino acids, whose sequence is coded for by the DNA of the genes, and this sequence determines how proteins will fold and then function. We age because of wear and tear, in a way not dissimilar to that of any machine, for example a car; death rates for cars follow a similar pattern to those for animals. There is no single ageing process. Ageing results from an accumulation of cellular damage and the limitations in the cells' ability to repair the damage, particularly in our DNA and proteins, and so restore normal function to the cell. The maintenance of the integrity of

DNA is a challenge to every cell, for such damage leads to the absence of key proteins, the synthesis of proteins in the wrong cells at the wrong time, and also to proteins with bad properties. Such damage accumulates randomly throughout life, from the time when body cells and tissue first begin to form. It is striking how organisms with the same genes, like identical twins, can age quite differently because of the random nature of the causes of the damage. Chance events are an integral part of ageing.

Cells are very complex and there are at least 150 different proteins that are involved in repairing DNA when it is damaged. Other damage occurs in mitochondria that produce the energy for cellular activities, and in membranes that surround the cell and are also present internally. How long we, and other animals, can live is determined primarily through mechanisms that have evolved to regulate the levels of cellular damage in the body. As discussed earlier, it is only the germ cells which give rise to eggs and sperm for reproduction that do not age as the damage is repaired. Because our germ-line – the cells that give rise to eggs and sperm – gives rise to the next generation it must avoid any damage due to ageing. This requires elevated levels of maintenance and repair in germ cells, as compared with body cells. Some trees can live 5,000 years, the reason being that there is no clear difference between germ and body cells, so there are mechanisms to prevent ageing in all their cells.

How do body-cell repair processes deal with the chemical diversity of the molecular damage that is central to ageing? The forms that damaged molecules in the cell and environmental toxins can take are almost limitless. It is another example of the brilliance of evolution that a set of genes has evolved to code for proteins that deal with the near-infinite structural diversity of molecular junk that accumulates with age. The most common molecular sign of ageing in cells is an accumulation of altered proteins derived from erroneous synthesis and wrong folding. There are a number of special proteins which help cells deal with proteins that have folded wrongly and other faulty proteins, and which can delay ageing and extend lifespan in some organisms. Protein turnover is essential to preserve cell function by removing proteins that are damaged or redundant. There is evidence that an accumulation of altered proteins contributes to a range of age-related disorders, such as Alzheimer's and Parkinson's disease. People with two copies of the longevity variant of the CETP gene involved in lipid metabolism have been shown to have slower memory decline and a lower risk for developing dementia and Alzheimer's disease

Cells can deal with the accumulation of damaged proteins and mitochondria due to ageing by eating bits of themselves – autophagy, the degradation of a cell's own damaged components. Autophagy can destroy damaged cell structures like mitochondria, cell membranes and proteins, and the failure of autophagy is thought to be one of the main reasons for the accumulation of cell damage and ageing. During ageing, the efficiency of autophagy declines, and damaged cellular products accumulate. TOR (Target of Rapamycin) is a protein enzyme which controls metabolism and can stimulate cell growth but can also block autophagy. Inhibition of TOR by rapamycin can increase lifespan in model organisms. In the nematode worm there is clear evidence that lifespan is linked to the capacity to regulate autophagy. Results from the fly demonstrate that promoting expression of an autophagy gene in the nervous system extends lifespan by 50 per cent, thereby providing evidence that the autophagy pathway regulates the rate at which the tissues age. Recent studies have revealed that the same signalling factors regulate both ageing and autophagy, and this involves longevity factors like the sirtuins which were discovered in yeast.

Although ageing is a multifactorial process with many mechanisms contributing, anything that damages the DNA and so leads to absence of proteins or faulty proteins can cause a malfunction of the cells. Some of the most important mechanisms causing ageing may involve damage to DNA. Damage to DNA can cause a mutation which alters the coding for a protein. There is also DNA in mitochondria that produce the energy for the cell. DNA may be the structure whose integrity cells have the most difficulty in maintaining over their lifetime. The DNA in every chromosome experiences thousands of chemical modifications every day and there is often repair – removal of damaged bases in the DNA is estimated to occur 20,000 times a day in each body cell. While DNA damage has not been shown to cause ageing directly, a number of rare human disorders, caused by mutations in DNA repair genes, include symptoms of premature ageing.

Cells tend to respond to serious DNA damage by committing suicide – apoptosis – and this provides a way of preventing the damaged cell becoming cancerous. This occurs much more often in aged tissues in which the background accumulation of damage is greater, and the resulting loss of cells may itself accelerate ageing. Long-lived organisms probably invest in better DNA maintenance. The benefit of this is seen both in slower ageing and delayed incidence of cancer, since genome instability contributes to both these processes. Humans are less likely to get cancer than mice, as they have invested more in DNA repair. While long-lived organisms make greater investments in cellular maintenance and repair than short-lived organisms, with age the repair mechanisms fade.

Nature and evolution seem to have a fine sense of irony when they made our lives so dependent on oxygen, which is essential for energy production but may also be a major cause of ageing and our eventual death. One possible cause of the damage to DNA and other molecules that leads to ageing places much of the blame on small modified oxygen molecules. Oxygen is required by the mitochondria in cells to produce energy from the molecules derived from food. The production of ATP, the key energy source in cells, by the mitochondria results in the production of these reactive oxygen molecules. Free radicals like reactive oxygen are formed due to loss of an electron which they steal from another molecule, and which makes them unstable and able to damage other molecules. Severe reduction of mitochondrial function in worms shortens lifespan significantly, and a prime candidate has been damage caused by reactive oxygen. Certain dwarf mice live almost twice as long and this is due to reduction in damage to the mitochondria in their brains. Many long-lived mutants are resistant to oxidative stress, and species of mammals that live longer tend to have cells that, when tested in culture, are more oxidative stress resistant. Large animals produce reactive oxygen at a slower rate.

Even single-cell organisms like bacteria and yeast age. The critical requirement for ageing in unicellular organisms is that a parent cell, when it divides, provides a smaller, and essentially younger, offspring cell. This occurs in yeast and the simple bacterium *E. coli*, and they share having a visibly asymmetric division and an identifiable juvenile phase. *E. coli* divides down the middle, giving each daughter cell one newly regenerated tip. But the cell's other tip is passed down from its mother, or grandmother, or some older ancestor. The cell that inherits the old tip exhibits a diminished growth rate, decreased offspring production, and an increased incidence of death. Thus the two apparently identical cells produced during cell division are in fact functionally asymmetric; the old cell should be considered an ageing parent repeatedly producing rejuvenated offspring. Asymmetric division may be a way for the cells to get rid of damage by dumping it into the older cell at division.

Yeast cells reproduce by the daughter cell budding off from the mother cell. After budding off some twenty daughter cells, the mother cell dies from what can be considered to be old age. Initially the mother cell buds every hour or so, but then the time between buds increases to three to four hours. Different strains of yeast age to different extents and the genes involved have been identified – the sirtuin (Silent Information Regulatory Two) genes. These genes are involved in life extension in a number of model organisms. In the nematode increased dosage of a sirtuin increases the mean lifespan by up to 50 per cent and involves the insulin signalling pathway. In flies, a sirtuin has also been reported to extend lifespan.

Model animal organisms have been invaluable in investigating what determines ageing and lifespan. These organisms include the nematode worm *C. Elegans*, which has about half our number of genes, a fixed small number of cells – 959 – and normally only lives for about 25 days; the fruit fly *Drosophila*, with an average lifespan of 30 days, which is a key model for genetic studies; and mice, which live for several years. The reason the nematode worm begins to die after a couple of weeks is due to the degeneration of its muscle after 15 days. Just why this occurs so soon is not understood, but the worm does not build its muscle nearly as robustly as mammalian muscle, and it contains no satellite cells that can replace damaged muscle cells.

Recent landmark molecular genetic studies have identified an evolutionarily conserved insulin-like growth-factor pathway that regulates lifespan in the nematode, fruit fly, rodents, and probably in humans. Reduction of the activity of this pathway appears to increase lifespan and enhance resistance to environmental stress. Genetic variation within the FOXO3A gene (the names given to genes can be quite weird), which can reduce this pathway's activity, is strongly associated with human longevity.

A dramatic example of an increase in lifespan came from the nematode worm. If the worms are placed under conditions where there is a limited food supply and many other worms are present, then instead of developing into adult worms through a series of larval stages, they develop into an alternative larval form known as a dauer larva. These dauer larvae neither feed nor reproduce, but if conditions improve they moult into adulthood and can then reproduce. But the dauer larvae, with their very dull lives, can live for up to 60 days, more than twice as long as normal worms. This is due to interference with the insulin pathway. Insulin plays a major role in the ageing process. A major discovery was a mutation in a single gene that caused the worms to live twice as long and remain healthy. This gene codes for a receptor for an insulin-like growth factor. The mechanism by which this increases longevity is not clear, but involves many other proteins. When sirtuins are over-expressed there is an increase in lifespan, and they were shown to interact with proteins of the insulin signalling cascade.

Reduced signalling by chemicals similar to our insulin also extends the lifespan of the fly *Drosophila*. It has recently been shown that in mice, less insulin receptor signalling throughout the body, or just in the brain, extends lifespan up to 18 per cent. Taken collectively, these genetic models indicate that diminished insulin-like growth-factor signalling may play a central role in the determination of mammalian lifespan by conferring resistance to internal and external stressors. The effects of eating less – calorie restriction – which can increase lifespan, also operate via the insulin effect. Fasting does reduce insulin secretion, but one must be cautious in trying too hard to reduce insulin secretion, as this can lead to diabetes.

There are genes that can extend lifespan or reduce it. The AGE-1 gene, for example, encodes part of a cellular signalling pathway that regulates dauer formation in the nematode worm via insulin-like growth-factor signalling. Mutations in genes encoding constituents of this pathway can extend lifespan not only in the nematode, but also in the fruit fly and the mouse. Single-gene mutations that affect longevity act via their interaction with multiple target genes. The increased lifespan in age-1 and related mutants in the nematode is associated with reduced reproductive fitness. The age of first reproduction is sometimes delayed or even prevented by the inappropriate formation of a dauer larva.

Sirtuins are also involved in mammalian ageing. A protein in the cell nucleus of mammals, NF-kappaB, is not only the master regulator of immune system responses, but can also regulate ageing. Activation of NF-kappaB signalling has the capacity to induce ageing in cells. Several longevity genes, such as the sirtuins, can suppress NF-kappaB signalling, and in this way delay the ageing process and extend lifespan. The protein SIRT1 – the mammalian equivalent of sirtuins – manages the packaging of DNA into chromosomes, and this role controls gene activity. When DNA damage occurs, SIRT1 abandons this critical task in favour of assisting with DNA repair. Mice that were bred for increased SIRT1 activity demonstrated an improved capacity to repair DNA and to help prevent undesirable changes in gene expression with ageing. It is involved in life extension that comes from calorie restriction.

There are other ways in which cells can age. A limit to the number of times some cells can divide in culture was discovered by Leonard Hayflick in 1965, when he demonstrated that normal human body cells in a cell culture divide about 52 times, but the number is less when the cells are taken from older individuals. There is no such limit for germ cells or cancer cells or embryonic stem cells. The explanation for the decline in cellular division of body cells in culture with age appears to be linked to the fact that the telomeres, from the Greek word for 'end part', which protect the ends of chromosomes, get progressively shorter as cells divide. This is due to the absence of the enzyme telomerase, which makes the telomere grow back to its normal length after each division. This enzyme is normally expressed only in germ cells, in the testis and ovary, and in certain adult stem cells such as those that replace cells in the skin and gut, as these cells have to be prevented from ageing. If the telomeres get very short, the cell is no longer able to divide and this means it cannot become a cancer cell. It may be that the telomeres can count how many divisions the cell has gone through, as they get a little shorter at each division. This could function to protect the cell against runaway cell divisions as happens in cancer, and ageing of the cells so that they have a limited number of divisions could be the price we have to pay for this protection.

Individuals can have their own telomere profile. In addition to the common profile, it is found that each person has specific characteristics, which are also conserved throughout life. Studies on both twins and families indicate that these individual characteristics are at least partly inherited. The length of individual telomeres might occasionally play a role in the heritability of lifespan. In diseases that result in premature ageing there is accelerated telomere shortening, and this may be partly responsible for the condition. There is new evidence that telomere shortening affects ageing in the general population, and is also likely to affect the way a person ages facially. A mutation in the so-called Peter Pan gene speeds up ageing due to telomere shortening. Up to 7 per cent of the population have two copies of this mutation, and they look up to eight years older than other people of the same age. About one third of the population has one copy, ageing them by three to four years. A fortunate, and fresh-faced, 55 per cent do not have the mutation and they remain youthful-looking for longer. Previous research has linked long telomeres with good health and shorter ones with age-related ills such as heart disease and some cancers. Shorter telomeres may thus be associated with shorter lives. One study found that among people older than 60, those with shorter telomeres were three times more likely to die from heart disease and eight times more likely to die from infectious disease. A study of centenarians, Ashkenazi Jews, found that their offspring have longer telomeres, and these are associated with protection from ageing diseases and better cognitive function, and can confer exceptional longevity.

There is increasing evidence that the nervous system may act as a central regulator of ageing by coordinating the physiology of body tissues. In worms, a number of different mutations that disrupt the function of sensory neurons extend lifespan. Furthermore, killing of specific neurons can increase lifespan in worms and flies. An intriguing question is whether functional disconnection in the brain leads to disruption of brain-systemic feedback loops involving crucial hormonal and autonomic systems. Such a loss of integrated function may contribute to age-related physiological changes, such as hypertension and insulin resistance, and predispose individuals to age-related pathological changes in the brain. It will be exciting to explore the extent of these functional connections in future studies.

It is an amazing fact that our skin cells are replaced about every 5 weeks, so by the time you are 20 years old you would have replaced your skin cells about 200 times. Do the cells that give rise to skin – skin stem cells – not age?

Stem cells are cells that divide and then one daughter cell remains a stem cell to divide again, whereas the other can become specialised cell such as a skin cell. Using mouse skin cells as a model system, researchers compared several properties of young and old adult skin stem cells. They found that, over an average mouse's lifetime, there was no measurable loss in their functional capacity. It seems that skin stem cells resist cellular ageing. There is no evidence that the lifespan of any species is determined by a limited supply or limited functionality of its stem-cell populations. An analysis of changes in gene activity as a mouse ages found that some tissues displayed large differences: in old mice, for example, there were genes in the brain which were more active, while other genes were less so, compared to a younger mouse.

Just less than one third of the variation in human lifespan is due to genetic differences that are important for survival after 60. Research on Danish twins born since 1870 found no evidence for an innate maximum lifespan shared by identical twins. Only about 25 per cent of the variation in adult lifespans could be attributed to genetic variation among individuals. The search for the genes that positively affect human ageing has been intense, but it has been very difficult. One example is the Peter Pan gene that extends human lifespan and acts via the insulin pathway which is so central in animal studies. Most of the long-lived men – those who eventually reached an average age of 98 years – had the same version of a gene which regulates the insulin pathway.

There are several illnesses related to old age which have a clear genetic basis and result in premature ageing. A genetic disease that causes premature ageing is Werner syndrome, which is a mutation in a gene that codes for a protein that unwinds DNA. Those with the illness typically grow and develop normally until they reach puberty, but usually do not have a growth spurt, resulting in short stature. The characteristic aged appearance of individuals with Werner syndrome typically begins to develop when they are in their 20s and includes greying and loss of hair; a hoarse voice; and thin, hardened skin. Affected individuals may then develop disorders such as cataracts, skin ulcers, type 2 diabetes, diminished fertility, severe hardening of the arteries, thinning of the bones and some types of cancer. People with Werner syndrome usually live into their late 40s or early 50s. The most common causes of death are cancer and atherosclerosis.

Premature ageing is known as progeria. Hutchinson-Gilford Progeria Syndrome – a very rare, genetic disease: only about 50 cases are currently identified worldwide – is due to a mutation in the LMNA gene that codes for a protein involved in the structure of the cell nucleus. Children with this mutation have small, fragile bodies, like those of elderly people, and typically live only about 13 years, and die from atherosclerosis and cardiovascular problems, although some have been known to live into their late teens and early 20s. The condition almost always occurs in people with no history of the disorder in their family. Whether the illness is similar to normal ageing is not known. Other genes are involved in age-related illnesses like Alzheimer's.

All these results suggest that no life strategy is immune to the effects of ageing, and therefore immortality may be either too costly or mechanistically impossible in natural organisms. Yet there are exceptions. Germ cells are immortal and a few primitive organisms, including hydra, a primitive simple animal in the form of a tube with tentacles, exhibit very slow or negligible ageing. Individual hydra were observed over a period of four years and yet showed no age-related deterioration, either in terms of survival or reproduction rates. The reason is not clear, but may be related to the fact that hydra can reproduce by forming buds which will develop into mature hydra without sexual involvement, and are also capable of undergoing complete regeneration from almost any part of their body. Most of their body cells can contribute to regeneration, so if some age, they may die or be lost during growth or budding.

Amongst the environmental factors that are linked to ageing, nutrition plays a prominent role. The great increase of non-insulin-dependent diabetes – type 2 – in industrialised nations as a consequence of a eating too much is an expression of this environmental challenge that also affects ageing processes. The most consistent effects of the environmental factors that slow down ageing – from simple organisms to rodents and primates – have been observed for calorie restriction. In yeast, the fruit fly and the nematode, sirtuins have been observed to mediate as 'molecular sensors' in the effects of calorie restriction on ageing processes. Sirtuins are activated when cell energy status is low.

Exposure to a variety of mild stressors such as calorie restriction and heat can induce an adaptive response that increases lifespan. For example, long-lived nematode insulin-signalling mutants are more resistant to thermal and oxidative stress. The term hormesis describes such effects, which are beneficial at a low level but harmful at a higher level. If induction of stress resistance increases lifespan and hormesis induces stress resistance, can hormesis result in increased lifespan? Here the answer is definitively yes. For example, in nematodes, brief thermal stress sufficient to induce tolerance to heat also causes small but statistically significant increases in lifespan. One possibility raised by studies of hormesis is that the increase in lifespan in animals due to dietary restriction, or to insulin signalling mutants, results from hormesis.

Increased longevity can thus be associated with greater resistance to a range of stressors. This may result from the increased expression of genes contributing to cellular maintenance processes, thereby protecting against the molecular damage that causes ageing. Similarly, the physiological stress of exercise has an optimal point for developing muscle strength and improving cardiovascular health, beyond which detrimental effects can be experienced such as attrition of cartilage in joints, leading to arthritis. Another possible example here is alcohol consumption: relative to abstainers, moderate drinkers have reduced mortality risk, especially from coronary heart disease. However, it is not known whether this effect involves stress-response hormesis. The study of stress-response hormesis and the induction by stressors of biochemical processes that protect against stress is providing new insights into the mechanisms that protect against a range of pathological processes, including ageing.

There is a great deal of research into the cellular basis of ageing and the progress is impressive, but there is still a long way to go before we fully understand how cells get damaged with time and, more important, how they repair that damage. One area that may illuminate the repair mechanisms will be by understanding how germ cells are prevented from ageing.

Professor Tom Kirwood is a leading scientist in ageing who gave the Reith Lectures in 2001. I asked him how much do we understand about ageing?

> We have a pretty good general understanding as to why ageing happens and a broad thrust of the mechanisms, but in terms of what there is still to be learned and in terms of any intervention we are only at the beginning. The number of scientists working on ageing is tiny compared, for example, to those working on cancer. Extending life expectancy is one of humanity's greatest successes, as we have doubled it over the last two hundred years, and for the first 150 of those years it was done by preventing people dying young by getting rid of infections and advances in general sanitation, vaccines and so on. Until about 25 years ago that was thought to be the end of the story. But it is a great surprise that the increase in life expectancy has not slowed one jot as people are getting old in better shape, and there is a decline in death rate in older people. After all this success, should we now be tampering with the ageing process itself?
>
> This raises challenging questions. In my view it is perfectly OK to use science to increase lifespan provided the emphasis is on the quality of the years gained. Is immortality possible? At a theoretical level, yes. When I had, some years ago just finished my book on ageing, *Time of Our Lives*, I had an idea for a work of fiction, a short story, 'Miranda's Tale', where science has managed to indefinitely postpone the ageing process. It has to be a possibility as the germ line does not age, they have better repair mechanisms, and there is also elimination of less good cells. It would not be by taking a drug, but would require changing our genetic constitution. There are animals like hydra which do not age. But I do not think it a practical objective. It is science fiction and should stay there.

What did he feel about his own ageing?

I think ageing is a challenging process – I am just coming up to 60 so not yet much affected. I enjoy being alive so I want to become older and I enjoy talking with older people. One has to accept reduction in mobility. Many young people do not want the problems of old age, but when they reach that age they may enjoy a very full and active life. Most people including scientists still think that we are programmed to age, although the evidence is totally against it – it is part of the need to see a purpose.

8
Extending

'Every Man desires to live long; but no Man would be old'
– Jonathan Swift

How long can we, and should we, live? How to live for ever has been a compelling subject for a very long time. One of the earliest legends about immortality, and also one of the earliest written stories, is that of Gilgamesh, the Babylonian demigod, from around 2000 BC. When he aged and began to fear death, Gilgamesh was told that he could survive forever if only he could show that he could master sleep by not sleeping for seven days and nights. But try as he might, he failed to do this. The gods then told him that he could find a plant underwater that, if he ate it, would make him alone immortal. Gilgamesh found the plant, but was enjoying swimming so much that he left the plant on the shore while he continued swimming. A snake came along and ate the plant. The lesson to be learned was that ageing was unavoidable.

The legends of the Greeks are filled with the adventures of the immortal gods and humans who seek immortality through their deeds, or through the acts of the gods. But there was also a more realistic attitude. The Greek philosopher Democritus criticised people for yearning for a long life, and argued that if they developed the right attitude to ageing and death they could live more peacefully.

The Roman Lucretius thought it absurd not to recognise that a long life was insignificant compared to how long one remains dead. He also argued that death was essential to keep the population down.

The ancients were all too aware of the dangers of immortality if the effects of ageing were ignored. This is illustrated by the story of Tithonus in Greek mythology, a story we should keep in mind when trying to extend life. Tithonus was the lover of the goddess of dawn, Aurora, and so good at what he did for her that she went to her father, the god of gods, Zeus, and asked if Tithonus could have eternal life. Zeus, being a doting father, immediately granted Tithonus immortality. The problem was that Aurora had failed to also ask for him to have eternal youth. With time the ageing process took its toll, and when Tithonus reached a hundred he had mild cognitive impairment and went around Aurora's castle babbling incessantly:

> But when loathsome old age pressed full upon him, and he could not move nor lift his limbs, this seemed to her in her heart the best counsel: she laid him in a room and put to the shining doors. There he babbles endlessly, and no more has strength at all, such as once he had in his supple limbs.

Aurora no longer loved him, and one day she turned him into a grasshopper, a cicada. Some claim that when today we hear the chirping of cicadas, it is just a group of old men babbling incessantly.

Judaism and Christianity had very strong views about long life. The Bible is against immortality; Psalm 90, verse 10, sets the human lifespan at threescore and ten, though the possibility of 120 years is given in Genesis. When Adam and Eve had eaten fruit from the tree of knowledge, God said, 'See! The man has become like one of us, knowing what is good and what is bad! Therefore, he must not be allowed to put out his hand to take fruit from the tree of life also, and thus eat of it and live forever.' God banished them from the Garden of Eden and stationed the cherubim and the fiery revolving sword to guard the way to the tree of life.

Old age could be the reward for a moral life and an indication of God's favour. 'Follow the whole instruction the Lord your God has commanded you, so that you may live, prosper, and have a long life in the land you will possess.' 'Ye shall harken diligently unto my commandments that your days may be multiplied, and that of your children,' says God in Deuteronomy. And in Proverbs: 'The fear of the Lord prolongeth days: but the years of the wicked shall be shortened.'

There are, nevertheless, claims in the Bible for very very long lives: Adam lives for 930, Noah 950 and Methuselah 969 years. Methuselah was a Hebrew patriarch and the grandfather of Noah; not much is known about him other than his extraordinary lifespan. There is also a figure from medieval Christian folklore whose legend began to spread in Europe in the thirteenth century, the case of Ahasverus, a Jewish cobbler, who told Christ to move on when he was carrying his cross and needed help. Christ told him that he would move on, but condemned Ahasverus to wander the earth and for his clothes to remain intact till Christ returned. And every 10 years he would be rejuvenated. There were reports that this Wandering Jew had been identified in 1252 at the Abbey of St Albans, and then again in Hamburg in 1642.

Many Indian fables and tales concern the ability to

jump into another body – performed by advanced Yogis in order to live a longer life. There are also entire Hindu sects, the Naths and the Aghoras, devoted to the attainment of physical immortality by various methods. Long before modern science made such speculation unreasonable, people wishing to escape death turned to the supernatural world for answers. Examples include Chinese Taoists and the medieval alchemists and their search for the Philosopher's Stone.

Denial of ageing can be very common, and longevity myths have been around for as long as humanity. Many of these legends involve places where people are reputed to live long. While it is true that people in certain cultures do not suffer many chronic illnesses while ageing, the lifespans of people in these places are hard to verify. As the *Guinness Book of World Records* stated in numerous editions from the 1960s to 1980s, 'No single subject is more obscured by vanity, deceit, falsehood, and deliberate fraud than the extremes of human longevity.'

There were claims that Thomas Parr was the oldest man who ever lived, surviving till 152 years. He was said to have been born in 1483 near Shrewsbury. He did not marry until he was 80 years old and had two children, a son and a daughter, both of whom died in infancy. He attributed his long life to his vegetarian diet and moral temperance, although when he was around a hundred years old he allegedly had an affair, and a child born out of wedlock. As news of his age spread, 'Old Parr' became a national celebrity and was painted by Rubens and Van Dyke. In 1635 he was brought to London to meet Charles I. In London he was treated as a spectacle, but the change in food and environment apparently caused his death. Charles I arranged for him to be buried in Westminster Abbey in 1635. William Harvey, the physician who discovered the circulation of the blood, performed a post-mortem on Parr's body. His autopsy suggested that Thomas Parr was under 70 years old.

In around 1500, Alvise Cornaro, an Italian nobleman aged 40, was feeling very unwell, and was advised by his doctor 'cut down on your riotous living, stop the drinking, cut out the rich food, eat as little as you can, and don't abuse your body. You can get well.' He wrote *The Art of Living Long* and argued that men and women were not destined to die at 60 or 70, but with care and a good constitution could live extremely long lives. The key to longevity lay in giving up excesses in all things, and he preached extreme moderation. He died aged 102.

Legends about healing waters abound. People have always talked about, and hoped for, water that will restore youth and health. Herodotus, the Greek historian who lived in the 5th century BC, provided much information concerning the nature of the world and the status of the sciences during his lifetime. He referred to a fountain containing a very special kind of water located in the land of the Ethiopians, and he attributed the exceptional longevity of the Ethiopians to this water. Tales of healing waters are also linked to Alexander the Great in his search for the Water of Life. Travelling almost to the edge of the world, Alexander finds a darkened country and travels in it with his servant Andreas. Alexander can't find his way through the darkness, but his servant does. Andreas drinks of the Water of Life and becomes immortal.

Another example that gave rise to the idea of a Fountain of Youth comes from the natives of Hispaniola, Puerto

Rico and Cuba, who told the early Spanish explorers that in Bimini, a land to the north, there were waters that had such miraculous curative powers that any old person who bathed in them would regain his youth. Juan Ponce de Leon, who had been with Columbus on his second voyage in 1493, and who had later conquered and become governor of Puerto Rico, is supposed to have learned of the fable from the Indians. The fable was not new, and probably Ponce de Leon was vaguely cognisant of the fact that such waters had been mentioned by medieval writers, and that Alexander the Great had searched for such waters in eastern Asia. Ponce de Leon, who had become wealthy in the colonial service, equipped three ships at his own expense and set out to find the mythical fountain that would restore his health and make him young again. What he found was not Bimini but Florida, and now many patients clearly believe he was successful, as during the winter months they eschew the care of their good doctors and flee south to Florida's warmer climate.

Myths about extreme old age persist into modern times. The Abkhasia are a people living in the Caucasus Mountains in southern Russia, a mountainous area near the Iranian border. They have a reputation for extremely long and healthy lives. In the 1960s and 1970s claims were made for lifespans of 150, marriages at 110 and fatherhood at 136. The greatest claim was that one man, Shirali Muslimov, was 168 years old when he died in 1873. The Soviets honoured him with a postage stamp. An official passport listed his birth date as 1805; Muslimov had no known birth certificate. The story was taken up by *National Geographic Magazine*, which later recanted on the claim. Individuals claiming to be physically immortal include Comte de Saint-Germain; in eighteenth-century France, he claimed to be centuries old, and people who adhere to the Ascended Master Teachings are convinced of his physical immortality. An Indian saint known as Vallalar claimed to have achieved immortality before disappearing for ever from a locked room in 1874.

Senescence results from a cumulative imbalance between damage and repair. Progress in reducing damage by improving living conditions and preventing disease, together with medical interventions, are fundamental causes of increased longevity. But myths apart, only around fifty people in human history have been verified as reaching the age of 114. Fewer than twenty of those who got to 114 have reached the age of 115. Worldwide there are estimated to be between 300 and 450 living supercentenarians – that is, over 110 years old – but as of June 2010 there is a list of only 79 validated supercentarians, and only three are male. Just how content they are is not clear.

Currently the oldest person is a Japanese lady Kama Chinen who is just short of 115 years. The title of the oldest verified person in history belongs to Frenchwoman Jeanne Calment (122 years and 164 days old), who died on 4 August 1997. She was born in Arles, France on 21 February 1875. Her genes may have contributed to her longevity as her father lived to the age of 94 and her mother to the age of 86. She rode a bicycle to the age of 100 and smoked till she was 117.

While the oldest woman was 122, the oldest man so far was 115. He was Christian Mortensen (1882–1998), a Danish-American whose age is undisputed, although the *Guinness Book of World Records* still ranks him second to the disputed case of Shigechiyo Izumi, 120, as the oldest man ever. Recently the oldest man alive, Henry Allingham, died at 113 in July 2009. There was much in the news about him; he had 6 grandchildren, 12 great grandchildren, 14 great-great grandchildren, and one great-great-great grandchild. He had had two mental breakdowns caused, he claimed, by working too hard. He lived in a home for blind ex-servicemen.

New evidence studying the genomes of 1055 centenarians found that it is now possible to predict if someone can live to 100 with a 77 per cent accuracy. The result is based on analysis of 150 mutations. It was found that 90 per cent of the centenarians possessed a particular genetic signature of mutations in the relevant genes. However, one must remember that this genetic test will not tell someone how long they *will* live, as genetics only accounts for about a quarter or a third of our lifespan; it could tell someone how long they *might* live.

The oldest known mothers are thought to be two Indian women, Rajo Devi and Omkari Panwar, who were allegedly both 70 when they had babies in 2008 following fertility treatment. But neither has a birth certificate to verify their age. The world's oldest father, Indian farmer Nanu Ram Jogi, fathered a child in 2007 at the age of 90. He is married to his fourth wife, boasts he does not want to stop, and plans to continue producing children until he is 100. Mr Jogi admits he is not certain how many children his series of four wives have borne him – but counts at least 12 sons and nine daughters and 20 grandchildren.

Alice Sommers was born in Prague 106 years ago and is the oldest person I know. She lives very close to me and is famed as a piano player and teacher – she still plays for several hours a day. She lives alone in quite good health, but goes out little, usually taken by her grandson. When asked how she felt about being old she replied:

> There are very good things. Experience. Looking backwards and enjoying knowledge. Only when we are so old can we appreciate the beauty of life. We are surrounded by miracles. Memories are so important. There are no bad things about growing old. None at all, and I am not at all afraid of death as that is the natural order of things. I was lucky to have been born with a very good temperament. When I am faced with a bad situation I immediately find something good in it. I do not think about how old I would like to get.

How is she cared for?

> I am looked after extraordinarily well – a girl comes in the morning for half an hour and then another for half an hour in the evening. And I get meals on wheels from the council. I use a magnifying glass to read so do not read much but Bach is my philosopher of music.

Those who pass beyond 90 do seem often to cope well, and centenarians can have daily lives that are as good as those ten years their junior. Being independent is a strong indicator for living long. In several studies, over one third of supercentenarians were still independent and able to care for themselves. In general it seems that this very old group are less well than the younger old, but then come to death rapidly. For example, only about 4 per cent die of cancer compared to 40 per cent of those around 50. They also have very low rates of heart disease, though there are stories that some are smoking heavily. There is a high incidence of some form of dementia, but not Alzheimer's, though their brains have the signs of that disease.

Of those who reach 100, a study found that about one half avoided chronic disease till they were over 80, and about one fifth escaped all the main chronic diseases. Children of centenarians suffer less from cancer and heart disease. A variation in the gene FOXO3A, a key regulator of the insulin-IGF1 signalling pathway, has a positive effect on the life expectancy of humans, and is found much more often in people living to a hundred and beyond – this appears to be true worldwide. The ApoE gene can also help with respect to dementia. Failing to give up smoking or to control blood pressure and cholesterol were reported to reduce life expectancy by 10 to 15 years. However, Clement Freud commented: 'If you resolve to give up smoking, drinking and loving, you don't actually live any longer, it just seems longer.'

Studies of twins and long-lived families have indicated that genes can explain about one third of maximum lifespan, but even identical twins age differently and this may be partly due to random switching on and off of some of their genes due to environmental influences. The other determinants are how one lives and chance factors like accidents and infections. Siblings of centenarians have a significantly higher chance of becoming a centenarian themselves. We have seen that the insulin IGF-1 system is involved in determining lifespan in model organism, so could reduction in its activity increase lifespan in humans? The answer seems to be yes. Mutations known to impair IGF-1 receptor function are overrepresented in a cohort of Ashkenazi Jewish centenarians, and DNA variants in the insulin receptor gene are linked to longevity in a number of groups located throughout the world. Increased activity of sirtuins, related to ageing in yeast, prompted by the drug resveratol has not been shown to extend lifespan in mammals.

In model organisms, such as the worm, fruit fly and mouse, changes in genes can dramatically increase their lifespan as much as fivefold. The equivalent life-extending effect in humans would result in an average lifespan of 400 years, and a maximum lifespan of over 600 years. But how healthy would such individuals be, and would they not have the effects of ageing? Never forget Tithonus. Many of the pathways regulating lifespan in model organisms are conserved throughout evolution, yet the genes that could dramatically increase human lifespan have not been identified.

How long would we like to live? Polls show that on average people want to live to about 90, though some 15 per cent had no idea how long they wished to live. Many were rightly very concerned about health as they aged, and one half, for example, feared the inability to drive their car. The elderly were less fearful than the young. Only about half of the public want scientists to work on mechanisms of age extension.

The Japanese artist Hokusai made his famous woodblock print of *The Great Wave off Kanagawa* in the 1820s at the age of around 65. Even after reaching the age of eighty, he was busy producing many fine prints. He often expressed his desire to live beyond the age of 90, and just before he died at the age of 88 he sighed and said his last words: 'If heaven gives me ten more years, or an extension of even five years, I shall surely become a true artist.' He also wrote:

> All I have produced before the age of seventy is not worth taking into account. At seventy-three I learned about the real structure of nature, of animals, trees, birds, fishes and insects. In consequence when I am eighty, I shall have made still more progress. At ninety I shall penetrate the mystery of things; at a hundred I shall certainly have reached a marvellous stage; and when I am a hundred and ten, everything I do, be it a dot or a line, will be alive. I beg those who live as long as I to see if I do not keep my word.

In *Back to Methusaleh* by George Bernard Shaw, immortality is important. The framing conception is that only the extreme longevity of Methuselah and other biblical patriarchs could provide humanity with the necessary wisdom for self-government. Shaw's solution is enhanced longevity: we must learn to live much longer. Yet earlier he wrote in *Misalliance*: 'After all, what man is capable of the insane self-conceit of believing that an eternity of himself would be tolerable even to himself?'

In Jonathan Swift's novel *Gulliver's Travels* (1726) the name Struldbrugg is given to those humans in a special nation who are born apparently normal but are in fact immortal. Although they do not die, they do nonetheless continue ageing. Swift's work depicts the evil of immortality without eternal youth, like the Tithonus legend. They are normal human beings until they reach the age of thirty, at which time they become dejected. Upon reaching the age of eighty they become legally dead, and suffer from many ailments including the loss of eyesight and hair:

> They were the most mortifying sight I ever beheld, and the women more horrible than the men . . . At ninety they lose their teeth and hair, they have at that age no distinction of taste, but eat and drink whatever they can get, without relish or appetite

> . . . In talking they forgot the common appellation of things, and the names of persons, even of those who are their nearest friends and relations . . . and whenever they see a funeral, they lament and repine that others have gone to a harbour of rest, to which they themselves never can hope to arrive. The reader will easily believe, that from what I had heard and seen, my keen appetite for perpetuity of life was much abated. I grew heartily ashamed of the pleasing visions I had formed, and thought no tyrant could invent a death into which I would not run with pleasure from such a life.

So if we wish to live longer we would like to be sure that we remain healthy and able to look after ourselves. Statistics show that currently the number of years a person can expect to live in poor health after the age of 65 is about five years. This means that somehow the bad effects of ageing must be prevented.

There have been many attempts to increase the length of life and avoid ageing. Towards the end of the nineteenth century the distinguished neurologist Charles-Édouard Brown-Séquard, at the age of 70 years, found that he was getting tired at night and introduced the first testicular-extract injections for rejuvenation. He advocated the hypodermic injection of a fluid prepared from the testicles of guinea pigs and dogs as a means of prolonging human life. This led to the Russian Serge Voronoff introducing 'monkey-gland' transplants to rejuvenate the ageing rich, as he believed that ageing was due to the slowing down of endocrinal secretions. In Kansas, John R. Brinkley's virility rejuvenation cure – transplanting goat gonads into ageing men – took the nation by storm. These are the historical precursors to the modern use of hormone replacement therapy, testosterone, for tiredness in males. A Swiss clinic from the 1930s injected organs from sheep embryos into patients' buttocks at a high cost; clients included, it is claimed, Churchill, Eisenhower and Pope Pius XII, but there is no reliable evidence that it worked.

It is widely held that a restricted but good diet, containing vegetables, fish, and fruit, together with exercise and having a positive attitude about ageing, can contribute to healthy and lengthy ageing. 'Blue Zones' is a name given to places where people live longer, healthier lives; although many of the claims are exaggerated, in all of these regions elderly people are much more active and youthful as they follow the rules for healthy ageing. Blue zones include Okinawa, in Japan, the most well-documented and studied population of centenarians, and the Hunza Valley in Pakistan. Legend has it that the Hunza people routinely live until 90 in good health, with many living as long as 120. They eat a diet primarily made up of fruits, grains and vegetables. The Vilcambamba in the southern region of Ecuador are reported to reach 100 and beyond, an achievement attributed to the natural mineral water.

A hormone secreted by the adrenal gland is half the level in men at age 60 compared to when they were 30. This reduction could be responsible for some of the ageing processes, but attempt to raise the levels in the old have not had useful effects. A number of other hormones, including growth hormone, testosterone, oestrogen and progesterone, have been shown in clinical trials to ameliorate some of the physiological changes associated with human ageing. No hormone, however, has been proved to significantly slow, stop or reverse ageing. It is possible that low testosterone could be the cause of tiredness, depression and sexual weakness in old men. Growth hormone is produced by the pituitary gland – a small structure at the base of thc brain – that is necessary for childhood growth, and also helps maintain tissues and organs throughout life. Beginning in middle age, the pituitary gland slowly reduces the amount of growth hormone it produces. Some reports suggest that growth-hormone treatment of elderly individuals can lessen some of the negative physiological changes observed with advancing age, but the results in humans have been controversial. Advertisements claim that by supporting the body's own natural production of growth nutrients, a human growth hormone such as Sytropin can have dramatic effects: 'If you are not completely satisfied, simply return for a full refund! It can help you maintain vigour, energy, and youthful enthusiasm. It provides a natural source of essential growth nutrients to assist the body's ability to withstand some of the harsh effects of ageing.' There is little evidence for this, or that taking this supplement promotes either longevity or an increased quality of life.

Dame Linda Partridge is a geneticist who studies the biology and genetics of ageing. I asked her how well we are doing:

> I think we understand it a lot better than we did because we have at last found a way in through the mutations in single genes that can extend lifespan in laboratory animals. This is a major discovery that started with the nematode worm and the longer lived worms remained healthy. Will those genes extend our lifespan? Probably yes as these genes can extend the lifespan of mice, and looking at old humans it seems similar processes are involved, like the insulin signalling pathway. It is not so much the increase in lifespan, but they keep the animal healthy as it ages and it covers a very broad spectrum of age related diseases, and so it's not just that they do not get cancer or cardiovascular disease but

they also get less osteoporosis, their skin is better. This is not what we expected to see. And it is jolly good news.

We need to make sure that this does help people. Not enough money is going into ageing research. The public needs to understand the advances and that it is a very good area to fund.

A possible problem is the relationship between scientists and clinical scientists, because geriatrics is mainly concerned with primary care and it is not very popular with students and does not have a strong research base. This means that it will probably be introduced with some medical speciality like endocrinology, cardiology or neurology. Pharmaceutical companies will make the connection – drugs will be the way in. It is most unlikely that it will involve modifying the genome. And one may have to take drugs for rather a long time as one is looking for prevention and so people will start taking the drug when they are middle-aged, and then continue for the rest of their lives – so the drugs have to be very safe. Life expectancy will probably continue to increase by two and a half years a decade. Immortality by altering our genes is beyond the limits of my imagination.

There is no evidence yet that dietary restriction helps with respect to ageing. When I meet people who are trying it they seem unhappy – their body temperature drops 2 degrees. Exercise is enormously beneficial to older people and quite simple exercise can make a significant difference to general mobility. I run two to three times a week – about 4 miles. I have found my ageing to have some enormous benefits and one or two irritations. One is into a health lottery with ageing and it depends on how that goes. I have been lucky as I have done nothing to look after my health. I do not think as fast as I use to, but enjoy the benefits of being more relaxed.

Getting rid of all diseases would, it is claimed, increase lifespan by adding only some 15 to 20 years to the current 80. For much longer lifespans, the ageing process in the cells and tissues itself would have to be reduced. There have been attempts to move in that direction, but even if immortality is not possible, can old age at least be delayed? The secret of long life is not known, and may not exist, but exercise and a good and varied diet seems to help. A study of 17,000 men from Harvard university over 30 years found that moderate exercise increased life by one or two years. Exercising regularly keeps the individual slim and fit, reduces stress, and increases the cardiovascular capability.

Evidence from animals is that limiting food intake, just eating less, can significantly extend the lifespan. When rats are kept in the laboratory under pleasant conditions but with an intake of food such that after weaning they get 50 per cent less than their well-fed neighbours, they live about 40 per cent longer. The oldest rat with high food intake is around 1,000 days old but there are those on the restricted intake who get to 1,500 days – 50 per cent older. In female rats, the age at which the ability to reproduce is lost is extended from 18 months to 30 months. Vitamins and minerals must be included in the diet, but it does not matter if the reduced calories come from carbohydrates, proteins or fat.

Dietary restriction can increase mice longevity but can impair immunity and wound healing. Low intake of calories suppresses most of the diseases common in older animals such as cancer, high blood pressure and deterioration of the brain. If the restricted feeding regime is returned to full feeding, the ageing process then seems to be actually accelerated. Monkeys on a restricted diet were three times less likely to get age-related illnesses. At the end of the study half of those on a normal diet had died, while for those on a restricted diet it was only one in five.

Considerable effort is being devoted to understanding

the molecular events mediating lifespan extension by dietary restriction, and whether sirtuins are involved. Many more studies in relation to dietary restriction are required for humans as there are negative side effects. A possible way to slow ageing would be to reduce metabolism by blocking receptors for insulin and growth hormone.

Evidence that humans could also delay ageing by reducing calorie intake comes from Japanese island of Okinawa, where there are probably more centenarians per 100,000 than anywhere else in the world. The average adult food intake is, for cultural reasons, 20 per cent less than the Japanese average, and schoolchildren on Okinawa eat less than two thirds of that recommended in Japan.The death rates from stroke, heart disease and cancer are only about two thirds that for Japan as a whole, and the death rate for 60-year-olds is half the national average.

Humans practising calorie restriction would reduce their calorie intake by about one quarter. For example, a person who needs 2,000 daily calories for weight maintenance might eat 1,500 to 1,600 calories a day on a calorie restriction diet. Eating so little might greatly increase lifespan, but who would want to live on just 1,600 calories a day? A single hamburger has 1,200 calories. But with the increasing prevalence of obesity, insulin resistance and type 2 diabetes, interventions targeting weight reduction and glucose control should be emphasised. Recent studies on the effects of lowering low-density lipoprotein cholesterol levels have shown a substantial reduction in mortality from coronary heart disease and nonfatal myocardial infarction rates, with a persistent effect in patients older than 75. Subjects with exceptional longevity and their offspring have significantly greater high-density lipoprotein, which reflect their general health and cognitive function performance. Rich diets shorten lifespan not because of excess calories but for more complex reasons – there is a dietary imbalance between fecundity and lifespan, each being maximised at different optimal nutritional levels. Certain amino acids – the structures from which proteins are built – can shorten lifespan in the fly but increase fecundity. This is a complex issue that is not yet solved.

Avoiding dementia, particularly Alzheimer's, is a key problem if life is to be extended, as this afflicts one in twenty of those over 65. Researchers from the Institute of Psychiatry in London say that every extra year worked delays the onset of dementia by just over a month. So working until you are 70 instead of 65 is likely to give you an extra six Alzheimer's-free months. I am not sure that is enough of a benefit to warrant the additional effort, but extending your working life is not the only thing you can do to protect yourself. The research supports previous theories that keeping the mind active for as long as possible can help to postpone mental decline. In contrast to earlier studies, however, the researchers found that the quality or duration of men's education or the type of work they did had no impact on the age of onset of the disease.

A community of elderly people in New York with an average age of 77 were monitored for five and a half years. Standard neurological and psychological tests for Alzheimer's were undertaken every 18 months. Higher physical activity was found to reduce the risk of Alzheimer's by a third, while people who had a Mediterranean-style diet rich in fruits, vegetables, cereal and fish, but low in meat and poultry, showed a 40 per cent risk reduction. Participants who both exercised a lot and ate a Mediterranean-style diet had a 60 per cent reduced risk. So probably the best way to reduce the risk of dementia is to combine keeping physically active with eating a balanced diet, and getting blood pressure and cholesterol checked regularly.

Another study found that adults who ate fish a few days per week were almost 20 per cent less likely to develop dementia than those who ate no fish at all. Although drinking coffee had previously been linked to a lower risk of developing Alzheimer's, a more recent study suggests that caffeine can directly target the disease itself. Mice with a rodent equivalent of the disease showed a 50 per cent reduction in levels of amyloid protein in their brains after scientists spiked their drinking water with caffeine. A small amount of alcohol can also help, but 10 per cent of cases of dementia are due to too much drink.

Attitude to ageing can itself have an effect and a positive attitude can extend life by some five years. In 1968 a team studied a group of people, who were aged 18 to 49, who completed a questionnaire that measured the extent to which they agreed with 16 negative views of ageing. These included beliefs that elderly people are 'feeble' and 'helpless'. Thirty years later, 25 per cent of those who had negative beliefs about ageing had suffered heart disease or a stroke, compared with 13 per cent of those who rejected the such views. Those who viewed ageing as a positive experience lived an average of seven and a half years longer. Women who were optimistic about their future were 14 per cent less likely to die from any cause than pessimists, and 30 per cent less likely to die from heart disease after eight years of follow-up in the study. Further evidence for the positive role of mental activity comes from a study of nuns who had long lives, even though they had plaques of amyloid characteristic of Alzheimer's disease in their brains. Positive self-perceptions of ageing have a greater impact on survival than lowered blood pressure or cholesterol. On the other hand, there is some evidence that stress, particularly short-term stress, can be beneficial and help reduce the ageing process, including Alzheimer's. A very provocative contemporary view is that laziness and lack of exercise and ambition will extend the lifespan.

Though thousands of years old the ancient Taoist tradition of eating mushrooms and other magical substances together with eating less, as some wandering monks practised in India to extend life, is still with us. There are modern religious mystics who believe in the possibility of achieving physical immortality through spiritual transformation as a part of their religious doctrines. They believe that after God has called the Day of Judgment, they will go to what they describe as Mount Zion in Africa to live in freedom for ever. They avoid the term 'everlasting life' and deliberately use 'ever-living' instead. An example is the Rastafarian and Jamaican singer Bob Marley, who refused to write a will despite suffering from the final stages of an advanced metastasised cancer on the grounds that writing a will would mean he was 'giving in to death' and forgoing his chance of living for ever. A group called the Rebirthers believe that they can acquire immortality by following the 'connected' breathing process of rebirthing.

The multi-million-pound industry based on anti-ageing treatments is discussed in the next chapter. In

2002 an article in *Scientific American* supported by some fifty scientists stated that the more dramatic claims made by those who advocate anti-ageing medicine in the form of specific drugs, vitamin cocktails or esoteric hormone mixtures are not supported by scientific evidence, and it is difficult to avoid the conclusion that these claims are made for commercial reasons. There has been a resurgence and proliferation of healthcare providers and entrepreneurs who are promoting anti-ageing products and lifestyle changes that they claim will slow, stop or reverse the processes of ageing.

The American Academy of Anti-Aging Medicine promotes the field of anti-ageing medicine and trains and certifies physicians in this speciality. Their co-founder Ronald Klatz stated that 'We're not about growing old gracefully. We're about never growing old . . . The leaders of the Anti-Aging movement will help to usher in a new modern age for humanity: The Ageless Society. There is a remedy for this apocalypse of aging, and this remedy comes just in time to save America.' But there is scientific hostility to its practices and no evidence that what they promote works.

Medical interventions for age-related diseases do result in an increase in life expectancy, but none have been proved to modify the underlying processes of ageing. At present there is no such thing as an anti-ageing intervention. For example at present there is relatively little evidence from human studies that supplements containing antioxidants lead to a reduction in the rate of ageing. The use of cosmetics, cosmetic surgery, hair dyes and similar means for covering up manifestations of ageing may be effective in masking age changes, but they do not slow, stop or reverse ageing.

Nevertheless, there is extensive advertising of anti-ageing products and the public is spending vast sums of money on them, even though in most cases there is little or no scientific basis for their promises and some may have harmful side effects. Some scientists are unwittingly contributing to the proliferation of these pseudoscientific anti-ageing products by failing to participate in the public dialogue about the reliable science of ageing research. There are, for example, advertisements for a treatment which prevents shortening of telomeres and so promotes longevity. Although telomere shortening may play a role in limiting cellular lifespan, there is no evidence that telomere shortening plays a major role in the determination of human longevity.

There are a variety of reports of substances that can extend life. A study of over a thousand men in Holland over 40 years found that those who drank half a glass of wine a day lived about five years longer than those who drank no alcohol at all, and two and a half years longer than those who drank beer and spirits. Herbs such as ginseng, rhodiola and maca have active ingredients that are claimed to suppress ageing. Studies show that a plant compound, resveratrol, can extend the lifespan of yeast, worms, flies and fish but there is as yet no evidence that it helps with humans. Resveratrol appears to mediate ageing effects partly by activating sirtuins. Resveratrol is found in the grape plant and in berries, and it is also a vital component of red wine.

A new star has appeared in the field of drugs that delay ageing in laboratory animals, and are therefore candidates for doing the same in people. The drug is rapamycin, already discussed in relation to TOR, its target, and which is in use for suppressing the immune system in transplant patients and for treating certain cancers. It can increase the lifespan of nematodes and fruit flies, and recently increased the lifespan of mice significantly. Given to the mice when they were 600 days old, it increased their lifespan by about 30 per cent. It has not been tested on humans and this should be done with great care because of its effects on the immune system. Studies in mouse models indicate that weakening the pathway on which rapamycin acts leads to widespread protection from an array of age-related diseases.

Aubrey de Grey is a scientist who, contrary to the standard scientific view, believes it will be possible to significantly prevent ageing. He calculates that two thirds of the people who die each day worldwide die of ageing, based on a definition of 'death from ageing' as death from causes that afflict the elderly more than young adults. He claims it will be possible to reduce the effects of ageing so greatly that humans will have a 50/50 chance over the next thirty years of being effectively immortal. He believes regenerative medicine may be able to thwart the ageing process altogether within that time. He works on the development of what he has termed 'Strategies for Engineered Negligible Senescence', a tissue-repair strategy intended to rejuvenate the human body. One basis for his claim is that mitochondria are damaged due to free radicals damaging their DNA, and they cause their host cells to secrete more damaging free radicals and so damage other cells. He believes that it will be possible to obviate the damage in the mitochondria's DNA. He also claims that many age-related degenerative diseases are linked to inadequate lysosomal function. Lysosomes are small vesicles in cells whose contents can destroy almost any unwanted cellular material – but not quite all, and this shortfall is known to underlie various age-related problems, including cardiovascular disease and macular degeneration. Alzheimer's disease involves the failure of unwanted proteins to be destroyed by other waste disposers both inside and outside the cell. Amyloid protein aggregates, a possible cause of Alzheimer's, are made from a normal protein that has been altered. An anti-amyloid vaccine could be helpful as the immune system would destroy the aggregates. A clinical trial along these lines was stopped when one individual became very ill, but a new trial has now reached phase 3.

In order to prevent ageing in this way, it will be necessary to repair, or else render harmless, numerous types of accumulating molecular and cellular damage so that the age-related pathologies caused by excessive amounts of that damage are prevented. In various cases, this requires manipulating genes, which at present would need to be done in the fertilised egg. And then the researcher would have to wait more than a hundred years to see if that individual survives that long, and suffers no problems from the manipulation to the genes. This is a most unlikely scenario; it is far too risky, and few if any researchers would live to see if their treatment worked. Thus, as de Grey accepts, the comprehensive application of regenerative therapies to ageing within a few decades relies on the development of safe and highly effective somatic gene therapy, which currently remains a daunting prospect. Additionally, an effective panel of therapies must address cancer, extracellular damage causing pathologies such as heart disease, and viral and bacterial infections. Finally, it is also far from clear that preventing cellular ageing would prevent cognitive abnormalities such as dementia and depression occurring, and de Grey acknowledges our ignorance of such matters.

In the UK the life expectancy for men and women is now 77 and 82, and a young man in his 20s today is expected to live five years longer than a man in his 50s. There are estimates that one in eight UK citizens now aged 35 will live to over 100. Half of the children alive today in countries with high life expectancies may celebrate their 100th birthday. In the UK, with a population of 61 million, 400,000 are over 90 and there are more pensioners than children under 16. The small village of Montacute in Somerset has the highest life expectancy for men in Britain, possibly because many of them grow their own food and work hard at it. In the USA, male life expectancy is 75 and female is 81. There is expected to be an increase from 4 million to 20 million over-85s by 2050 in the USA, at which time there may be nearly one million centenarians. At present about 10 per cent of the world's population is over 60 but by 2050 it will be 20 per cent, and the elderly will outnumber children worldwide. Women outnumber men at age 100 by 5 to 1. Japan has the current highest life expectancy for females, estimated at 85, and Iceland for males, at 80 years.

There are currently about 40,000 centenarians in the United States, and they are the fastest growing segment of the population. Traditionally scientists believed that most people who live to 100 experience a 'compression of morbidity' – that is, they do not develop common age-related chronic illnesses like diabetes or coronary disease until very late in life, if at all. However, more recently, investigators have found that nearly one-third have in fact suffered from long-standing chronic illness, in many cases for 15 years or more, before turning 100. What they experience is a compression of disability: they avoid major disability and require little or no assistance in performing the activities of daily life, at least until extreme old age.

The increase in the numbers of elderly people throughout the developed world is already having serious consequences, which will only increase in the future. Leon Kass, who has been chairman of the US President's Council on Bioethics, has questioned whether the resulting over-population problems would make life extension unethical. 'Simply to covet a prolonged lifespan for ourselves is both a sign and a cause of our failure to open ourselves to procreation and to any higher purpose . . . Desire to prolong youthfulness is not only a childish desire to eat one's life and keep it; it is also an expression of a childish and narcissistic wish incompatible with devotion to posterity.' He highlighted the importance of lifespan limits in making room for new generations who deserve to take their rightful place in the world.

Francis Fukuyama, who predicted the eventual global triumph of political and economic liberalism, argues that efforts to increase human longevity risk undermining social security schemes, damaging family structures, and rendering the United States vulnerable to assault by countries with more youthful populations. He suggests that it could lead to a 'posthuman future', where human existence would be radically different from what we currently experience. This may be an alarmist view, but the problems associated with increasing human lifespans are still severe.

What impacts might a much further increase in age in the population have? Would immortality be a benefit or a disaster? In Kurt Vonnegut's 1998 story 'Tomorrow and tomorrow and tomorrow' a grandfather aged 172 takes an anti-ageing potion; he drives his descendants mad by taking the best food and space. Many fear that extending lifespan without reducing illnesses would increase the time living with limited physical and mental abilities, but it could nevertheless offer new opportunities. Superlongevity, radical life extension, would require every citizen to learn new skills. There would probably be a craving for novel experiences. Even now, many of those who have retired feel much younger and wish to have an active life that can include new work and learning. Living longer would enable people to find out what the future is like, but they would need to be healthy and cared for – and not be bored. And note again that in spite of all the research, no way of preventing ageing other than by a healthy lifestyle has yet been discovered.

Nevertheless there are several organisations devoted to extending lifespan and even superlongevity. Leon Kass apparently believes that if our bodies don't grow old we will become even more fearful of death. He also thinks we will feel unhinged and lack the sense of purpose that supposedly comes with growing old. Superlongevity would make time of no consequence and this could have bad consequences due to the increase in population size. The emerging picture of perhaps many hundreds, or even thousands, of small effects and tissue-specific damage provides a sobering challenge for those aiming to engineer reduced senescence. Controlling behavioural and environmental exposures to reduce cell damage may be a more realistic priority, as the great majority of people are likely to have large numbers of genetic vulnerabilities for one or another disease not related to ageing. A £5 million programme of bioengineering has been proposed to do research to find solutions to the problems associated with ageing of the body. Research will focus on joints, spine, teeth, heart and circulation. This seems more sensible than trying to develop safe and effective genetic engineering to alter the thousands of small damaged functions in our cells.

In Shakespeare's *All's Well that Ends Well* the king is suffering from physical disability related to old age. When Helena offers to cure him, which she later does with a potion from her doctor father, the king responds with cautionary words:

> We thank you, maiden;
> But may not be so credulous of cure,
> When our most learned doctors leave us and
> The congregated college have concluded
> That labouring art can never ransom nature
> From her inaidible estate.

9
Preventing

'To get back my youth I would do anything in the world, except take exercise, get up early, or be respectable' – Oscar Wilde

If we cannot be immortal, can at least our youthful looks be maintained? Almost everyone wants to look reasonably young while still living to a respectable old age without serious disabilities. Youth's attraction is no mystery: evolution wants us to reproduce and so has selected us to find young people attractive, since they are the best reproducers. The same principle has resulted in us finding old faces unattractive. What could we do to avoid changes in our appearance with age? Having the right genes is a good beginning, as is keeping fit and active, and eating the right foods – staying slim is one of the key factors to looking young, but it does not hide wrinkles.

Efforts to hide and prevent ageing are far from being a modern obsession. In Ancient Egypt cosmetics were applied to the face and eyes, and cosmetic implements, particularly eye-makeup palettes, have been discovered in the earliest graves. Honey as well as various herbs and plants were used in an attempt to devise anti-ageing treatments. The aloe plant was commonly used as an anti-wrinkle treatment and is still with us today. Cleopatra is known to have used lactic acid in order to peel her skin, believing it made her appear more beautiful. The arid desert climate of Egypt led to the widespread use of body oils as moisturisers. It is believed that all classes of Egyptian society were concerned with their appearance, both men and women.

This pattern is repeated throughout the ancient world. As now, the focus was on the youthful beauty of women rather than men. The Roman poet Ovid despaired of time's encroachments: 'The years will wear these charming features; this forehead, time withered, will be crossed with wrinkles; this beauty will become the prey of the pitiless old age which is creeping up silently step by step.' Other writers saw the comedy as well as the pathos of the situation. 'The Man and His Two Mistresses' is one of Aesop's *Fables*, written around 600 BC:

> A man of middle age, whose hair was turning grey, had two mistresses, an old woman and a young one. The elder of the two didn't like having a lover who looked so much younger than herself; so, whenever he came to see her, she used to pull the dark hairs out of his head to make him look old. The younger, on the other hand, didn't like him to look so much older than herself, and took every opportunity of pulling out the grey hairs, to make him look young. Between them, they left not a hair in his head, and he became perfectly bald.

In a recent survey many men and women said that they are, will be, or were, at their physical peak not during their youth but during their early middle years around the age of 40. Those aged 65 and over said 46 was their personal best age. But in terms of appearance, youth remains the golden age. Marie Helvin, at 54 still a supermodel, said: 'Please shoot me if I'm doing this in my 80s. Anyway, one day I won't be able to. My mother always said that Japanese women look youthful for years and then one morning they wake up and they've aged like 100 years. And she's right. It happened to her when she was 79.'

Celebrities and many others have fallen prey to the cloned-youth look. The American anti-ageing magazine *New Beauty* offers articles on how to get flawless feet, and lists the top ten wrinkle reducers. However, the treatment needed to achieve this youth has, it is claimed, made many women look like waxwork escapees from Madame Tussaud's. Many have had their faces injected with a filler to remove the creases while others have plastic surgery. In a survey 20 per cent of men said they thought that cosmetic surgery for their wives could save their marriage; it seems no one asked the women whether they would like the pot bellies of their husbands reduced.

Currently the global anti-ageing market for cosmetic products and treatment is estimated to be worth approximately $57 billion, a figure that is expected to grow at breakneck speed in coming years. In the UK cosmetic surgery has tripled in the last five years. Britons are spending nearly £500 million a year on cosmetic procedures, said *Which?* magazine, more than any other European country. A total of some £673 million a year is spent on skin care, and these figures are dwarfed by figures from the US. Of course not all this money is spent by the old, but anti-ageing products are the largest growing sector. An article in *Time* magazine in early 2009 introduced the concept of 'amortality' when referring to the current attempts to avoid ageing and achieve a leap in life expectancy. Age-appropriate behaviour, it claimed, will be relegated to the past, like black-and-white television. Amortals do not dread extinction – they deny it.

Surveys exploring attitudes towards ageing, beauty

and cosmetic surgery can yield varying results. In one survey of some 2,000 Americans aged over 18, as well as 500 who have had cosmetic surgery, almost all of those interviewed were satisfied with the way they look for their age, and over half felt that inner beauty is more important than physical appearance – this was particularly true for the old. Just one in three said physical beauty counts most. More than half believed that men and women age gracefully, and only a quarter of women felt that maintaining an attractive physical appearance was important for them. Most of the women were satisfied with their appearance. After 45, women were more interested in looking good for their age than in trying to look a different age. Agony Aunt Virginia Ironside has commented, 'I want to look good at my age, but I also want to look old enough for people to open heavy doors for me.' She also very much appreciates being offered a seat on a crowded bus.

But in another survey, almost three quarters of women cited body shape as a 'major concern'. Meanwhile, men are also taking more time over their appearance. About 20 per cent said they would consider getting cosmetic surgery in the future, while about 22 per cent were unsure if they would. Those under the age of 40 were nearly twice as likely to consider having a procedure in the future. A study by *Which?* magazine found that Botox treatments are seen as a desirable Christmas present by 50 per cent of people aged between 16 and 24, and 45 per cent of those aged 55 to 64. A survey of schoolchildren found that 18 per cent of all boys and 25 per cent of all girls declined to imagine any form of enhancement because they saw it as unnatural or simply unnecessary. One girl commented, 'I wouldn't want an upgrade because I wouldn't want to be different. I like being who I am.'

Despite people's professed opinions, however, cosmetic surgery in general has been growing in popularity in Britain, with a threefold increase in the first decade of the twenty-first century. According to figures released by the British Association of Aesthetic Plastic Surgeons, only 10,700 procedures were performed in 2003, but by 2009, this figure had risen to 36,482. One of the biggest growth areas was in people in middle or late middle age. These figures do not include non-surgical interventions like the use of Botox, which has been increasing even more rapidly. The most common procedures were breast enlargement, liposuction, and eyelid and facial surgery.

In the US, over 10 million surgical and non-surgical procedures were performed in 2008, costing over $11.8 billion. Men had over 800,000 cosmetic procedures. Strikingly, people aged between 54 and 61 had about a quarter of these procedures, and people over 65 much fewer. Over half were for breast augmentation and fat reduction. Plastic surgery of the face can help you feel better, but does not affect what is going on inside the body. Anti-ageing surgery procedures are widely advertised on the internet with the injunction to 'Get expert free advice'. These include Eye Bag Removal, which can restore a youthful look through the removal of fat and excess skin from both sets of eyelids, and the Brow Lift is a procedure that concentrates on restoration in the upper part of the face, correcting drooping eyebrows and loose skin in that area. Facelift surgery challenges the most visible signs of the ageing process: loose facial and neck skin is removed to produce a smoother, fresher appearance. The operation can take between two and three hours and it is recommended that patients spend at least one night in the cosmetic surgery clinic after the procedure. Liposuction entails the removal of fatty deposits from any part of the body, a process usually taking no more than one and a half hours to complete. The recovery time involved is minimal, with most patients getting on with their lives again within a few days.

Those undergoing or planning to undergo cosmetic surgery still constitute a distinct minority, but the numbers would increase if it could be done safely. As with any surgery, there are some risks associated with these procedures. And not every procedure will have the desired results, as the case of the unfortunate American Jocelyn Wildenstein, who allegedly spent $4 million on facial plastic surgery to please her husband, illustrates. She has been unkindly dubbed 'The Bride of Wildenstein', and her horrified husband apparently commented, 'She seems to think you can fix a face the same way you fix a house.' But the emotional benefits of plastic surgery results can be many times greater than the physical rewards. If you have felt bad about the way you look, plastic surgery can make you feel better about yourself.

Treatment based on Botox injections is probably the most popular cosmetic surgery worldwide. The adverts refer to the frown lines, crow's feet and wrinkles around the mouth that can cause a person to look worn out, tired and old. Botox treatments were first approved by the FDA in 1990 for the treatment of eye muscle spasms for those under 65; however, its cosmetic value was quickly realised. Botox cosmetic is an approved trade name for botulinum toxin, produced by the bacterium

Clostridium botulinum. When injected in small doses in designated areas, Botox blocks nerves activating muscles responsible for the repetitive action that causes fine lines and wrinkles, and literally paralyses the area. Normal Botox treatment with a doctor will cost about £400, and results can last up to eight months. But it doesn't always, of course, have the desired effect, and can lead to a face that looks angry as well as to headaches. One actress claimed that it made her look like an extra from the *Planet of the Apes*.

A multitude of wrinkle creams and lotions sold in chemists and department stores promise to reduce wrinkles and prevent or reverse damage caused by the sun. At the top end of the market, Oro Gold Cosmetics has introduced a set of anti-ageing products (including eye serums, bionic facial treatment and skin moisturisers and other products) infused with 24k gold – which, they claim, but without reliable evidence, has many properties that induce rejuvenation in the skin. But do any of these products actually work? Some research suggests that wrinkle creams do have ingredients that may diminish wrinkles. But according to Mayo Clinic physicians, many of these ingredients have not undergone scientific research to prove their benefit. Creams and lotions may slightly improve the look of your skin, depending on how long you use the product and the type and amount of the active ingredient, but any effects from non-prescription wrinkle creams will not last very long. You'll have to dab on wrinkle creams once or twice a day for many weeks before noticing any improvement. And once you stop, your skin will very likely return to its original wrinkled appearance, according to dermatologists. Studies have confirmed that more expensive wrinkle face creams work no better than cheaper products.

Which? magazine decided to put anti-ageing creams to the test. It selected 12 ordinary moisturisers and 12 anti-ageing creams. Groups of four women tested each product for four weeks. None of the 96 women knew which product they were using and at the end of the trial they were asked to guess whether they had been using a moisturiser or an anti-ageing cream. Three quarters chose moisturiser. Most had not noticed any difference in the look or feel of their skin, and of the 48 women who had been using an anti-ageing cream, only 10 reported any improvement.

Which? concluded that 'some of the claims made for the ingredients of anti-ageing creams can be substantiated but, with the low concentrations used in the creams, they are unlikely to do more than moisturise your skin'. There is more support for this view from a beauty-industry insider: 'There is no miracle ingredient that will take years off your appearance,' says Gisele Mir, a cosmetic scientist and founder of the holistic skincare range Mir. 'The only miracle is that the cosmetics industry has managed to persuade us otherwise for so long. In my opinion you can harm your skin by using anti-ageing products. I believe many of these products accelerate ageing rather than prevent it.' Tretinoin, a derivative of vitamin A, is the only topical medication that has been proven to improve wrinkles.

The advertisements and publicity seem endless. A full-page advertisement in several UK newspapers proclaims: 'Is your skin ageing too fast? Our scientists definitely think so . . . Inspired by 25 years of groundbreaking

DNA research, Estée Lauder now innovates anti-ageing skincare . . .' In the US: 'Discover the FIVE SECRET RITUALS from a Hidden Himalayan Monastery that Make You Look 30 Years Younger – In Just 10 MINUTES A DAY! $39. This unique complex of Bio Enhanced trans-resveratrol and potent polyphenols and anthocyanins is formulated specifically to fight ageing at both the genetic and metabolic level.' It is claimed that CoverGirl Simply Ageless Foundation significantly improves skin condition in just four weeks, and super model Christie Brinkley's affiliation with CoverGirl was a huge success. She said: 'CoverGirl is part of my DNA and I'm thrilled to be back with my family. I'm excited to promote a new product developed specifically for women like me, who want flawless coverage combined with the latest science in skincare. I'm looking forward to working with the brand at a time when there is so much innovation.' But there is as little real evidence for the benefits of agents that prevent facial ageing as there is for medical treatments.

In April 2009 there was a report that a new product from the chemist Boots, which works by stimulating the production of a protein that promotes skin elasticity, made a significant difference to the wrinkles of 70 per cent of volunteers aged between 45 and 80. This was a properly conducted trial by the University of Manchester and so seems an important advance. It led to a stampede to Boots' shops which is still continuing. In one month, February 2010, Boots sold more than 700,000 anti-ageing creams. However, other experts say that only one in five users will get something more than from using common moisturisers. A new product of a quite different nature is a device that electrically stimulates the muscles in the face; there are claims that it is safer and more effective than plastic surgery, and acts due to stimulation of the facial muscles, but there are no reports of proper trials.

The links between 'expert opinion' and the commercial interests behind a product are always worth considering. A recent report in the *Boston Globe* said that when a 73-year-old patient recently asked a prominent dermatologist how she could look more youthful, he had a ready answer. He suggested she visit a drugstore across the street to shop for a specific brand of anti-ageing, wrinkle-fighting, and lip-plumping creams. The products were part of the doctor's own line of cosmetics.

The use of wrinkle creams twice a day can plump up the skin temporarily, which causes the wrinkles to be less visible. These creams may also be excellent moisturisers, and they may smell and feel good. But no matter how expensive they are and no matter what they claim to do, they won't turn back the clock. An important way to avoid wrinkles is to practise sensible sun avoidance and use sun screen. Yet are wrinkles really such a terrible sign of ageing compared to certain other indicators?

Weight and other factors can affect how one looks. A study in the USA of the photographs of 186 pairs of identical female twins together with detailed interviews showed surprising results. One twin aged 70 could look six years younger than her sister if she had more weight, as this filled in the wrinkles on her face. Antidepressants and alcohol made a twin look older, as did suntan and cigarettes. Before 40, the lighter twin looked younger and more attractive, but over 40 the heavier twin looked younger, the weight difference being about 24 lbs. So it may well be that what you eat has a more beneficial effect than what you dab on your skin in the war against wrinkles. But being overweight, of course, can have many ageing effects on one's health.

There are studies that suggest that hair is more important than wrinkles in judging someone's age – it can make a differences of 4 to 5 years in one's appearance. Also important is the fullness of lips, which makes a face appear much younger and is almost entirely determined by a person's genes. This accounts for the large number of cosmetic surgery procedures for lips, some leading unfortunately to 'trout pout'.

It is reassuring to learn from the attitude surveys that many think real beauty comes from the inside, not the outside. And in spite of the enormous amounts of facial treatments to avoid looking old, there are those who argue that we should learn to live with this universal process. Anne Robinson, a well-known TV presenter in her 60s, had a facelift at 61 and currently uses Botox. An article in the *Daily Mail* recognises that it is tough to still be a TV presenter at her age; but would it not be better, it argues, for someone like Robinson, who is a role model for women, to champion the rights of women as they grow old? When women like her resort to cosmetic surgery it puts unwelcome pressure on women of a similar age. As a cosmetic surgeon comments, facelifts are dangerous, and he is dismayed that injectible fillers for wrinkles are advertised on TV.

All these procedures can lead to an obsessive desire to achieve an ideal beauty. An article in the London *Evening Standard* offered ten tips to make ladies look ten years younger: remove dead cells from your skin using a wrung-out face cloth every night; fill out the volume of your face with injections; get baby Botox to get rid of embedded frown lines; wear sunscreen every day; use an expensive serum to exfoliate; get rid of uneven skin tone using a light concealer; get made up by an expert; get advice on hair colour and shape; tint eyelashes and eyebrows; make sure your teeth are a harmonious shade of pale. How much time will be left in the day for life's other pleasures?

Oscar Wilde found a way to prevent the signs of ageing in *The Picture of Dorian Gray*. Dorian is a cultured, wealthy, and exceptionally good-looking young man who has his portrait painted. He curses his excellent and attractive portrait, which he believes will one day remind him of the looks he will have lost as he ages. In a fit of distress, he pledges his soul if only the painting could bear the burden of age and infamy, allowing him to stay young for ever. This wish is fulfilled and he does not show signs of ageing, but the portrait does. Only when he dies does the portrait turn back to a youthful image, while his body show all the wrinkles. A well-known quotation from the book says: 'The tragedy of old age is not that one is old, but that one is young.'

Perhaps Lucille Ball got it right: 'The secret to staying young is to live honestly, eat slowly, and lie about your age.'

10
Treating

> 'Age appears to be best in four things: old wood best to burn, old wine to drink, old friends to trust, and old authors to read'
> – Francis Bacon

Evolution cares not for the old once they no longer contribute to reproduction or the care of those who can reproduce. And while love for children is universal and genetically determined, since reproduction is what life is for, attitudes towards the old are not. How the old are treated can vary in different cultures; even within a single society people do not look upon the aged as belonging to one clearly defined category, and attitudes are diverse.

Largely as a result of increases in retirement age, and of people living and working longer, attitudes to ageing and being old have changed, and nobody really knows what 'old' is any more. A recent UK survey found that, on average, the public believed that 'youth' generally ends at 45 years of age, and that 'old age' starts at 63 years of age. Older respondents considered that youth continues longer and old age starts later than did younger respondents. The oldest age group of those in the survey thought that old age started at just over 70, whereas the youngest group estimated old age started at around 55. In terms of the end of youth, the oldest estimated age for this was 57, whereas the youngest was 37. Categorisation of 'old' and 'young' is so variable that an older person assumes someone is still in their youth at 57, whereas a younger person assumes that by this age they are already old. Ageing now happens more slowly, and people get 'old' later. This post-modern attitude to ageing reflects a feeling that while ageing comes to everyone, how you deal with it by keeping active, both physically and mentally, can put off 'getting older'. But there are many views that make those who are ageing, and particularly women, refuse to tell even friends their true age.

'Implicit ageism' is the term used to refer to the unconscious negative thoughts, feelings and behaviour relating to older people. Becca Levy, whose research explores psychosocial influences on ageing, focuses on how psychological factors, particularly older individuals' perceptions of ageing, affect cognition and health in old age, and reports that they 'tend to be mostly negative'. One can compare these attitudes with those of the Nambikwara Indians, who live in the south-western part of the Brazilian Amazon, and who have only one word for young and beautiful, and another for old and ugly. The old are on the whole viewed as physically unattractive. This makes sense, as since the old no longer reproduce, from an evolutionary point of view they have lost all beauty; but they can help children and younger people, which does not require attractiveness.

It is far from clear whether views of the old in ancient times have affected modern views. In ancient Greece, Sophocles, Euripides and Plato lived productively into their 70s, and the views they espoused of the elderly were positive and respectful. But we should remember that there were only a few old in those times, with life expectancy being around 30, and half of those born not passing the age of 10. The chief killers were infectious diseases such as typhoid, smallpox, cholera or malaria. So to reach the age of 80 was exceptional. Many Greeks thought physical decay with age a curse worse than death itself.

Plato and many of the Ancients had a positive view of old age: 'Old age has a great sense of peace and freedom. When the passions have lost their hold, you have escaped, as Sophocles says, not only from one mad master, but from many!' Plato also wrote that 'As age blunts one's enjoyment of physical pleasures, one's desires for the things of the intelligence and one's delight in them increase accordingly.' He emphasised the respect with which children should treat their parents, and both he and Socrates pointed out that one could learn much in the company of the elderly. In Sparta the old were protected and venerated, and government policy was made by a council of twenty-eight elders over the age of 60, elected for life. But there were other views.

Aristotle praised youth and his views of the old were quite the opposite to those of Plato: 'Because they have lived many years, because they have often been deceived, because they have made mistakes, and because human activities are usually bad, they have confidence in nothing and all their efforts are quite obviously far beneath what they ought to be.' For Aristotle, man only advanced until the age of 50, and when older became garrulous and kept on going over the past. A distaste and disgust for old age was openly expressed in Greek culture. Many believed that the gods took those they loved at a young age, leaving the unwanted to experience old age. Yet several Greek laws were passed requiring children to provide for their parents, and there were severe penalties – including, in Athens, the loss of civic office – for those found guilty of maltreating their parents.

In Ancient Greece, Aristophanes was among the first to mock the old in his plays as being feeble. Euripides also had a negative view of old age; in his play *Alcestis*, Admetus says 'Old people always say they long for death – their age crushes them – they have lived too long. All words! As soon as death comes near, not a single one wants to go, and age stops being a burden.' Nor was Aeschylus in the *Agamemnon* any more positive:

> What is an old man?
> His foliage withers
> He goes on three legs and
> No firmer than a child
> He wanders like a dream at noon.

The riddle of the Sphinx who guarded the entrance to Thebes is well known: what, she asked, has one voice and is four-footed, two-footed, and three-footed, and goes slowest when it has the most feet? Oedipus, passing by, answered: it is a human being that starts on all fours, is mature on two feet, and then when old has three, as there is also a cane. The Sphinx killed herself when, on this occasion, her riddle was answered correctly.

The philosopher Cicero, who introduced the Romans to Greek thought, was positive about ageing, celebrating the delights of intellectual activities in old age such as civic service, writing, learning a language, and the study of philosophy. But he also listed the difficulties:

> As I give thought to the matter, I find four causes for the apparent misery of old age: first, it withdraws us from active

> accomplishments; second, it renders the body less powerful; third, it deprives us of almost all forms of enjoyment; fourth, it stands not far from death.

The Roman poet Ovid was also unenthusiastic: 'Farewell to laughing, happy love and easy sleep', and 'Time, oh great destroyer, and envious of old age, together you bring all things to ruin.' It has been estimated that about 20 per cent of the senators in Rome at any one time would have been 60 years of age or older. Both Cicero and Plutarch, in their own old age, felt that their years did not earn them the respect they merited. The Roman playwright Plautus created sympathetic old male characters, and in one play points out that an old man should be careful to avoid prating about public affairs, or slipping a hand under the dress of a woman whom he does not know. It has been suggested that from Ancient Egypt to the Renaissance the theme of old age was handled by writers in a stereotyped manner, with similar comparisons being made and little attempt to really look at old age deeply.

Respect for the old was an important principle in Judaism. The Old Testament says that we must cherish parents in their old age: 'Thou shalt rise up before the hoary head, and honour the face of the old man, and fear thy God' (Leviticus 19:32). This has been taken to mean that when an old man or woman passes by, you should stand up as a token of respect. Old age may be one reward of those who honour their parents: 'Honour your father and your mother so that you may have a long life in the land that the Lord thy God giveth thee' (Exodus 20:12). The Koran takes a similar view: 'Be good to parents, whether one or both of them attains old age with thee . . . neither chide them, but speak unto them words respectful.' To

Buddha, born in 565 BC, old age was a spectacle of misery and sorrow which needed to be eliminated. Against this, the *Upanishads*, the sacred texts of Hinduism, speak of active and joyful ageing.

Since Jesus was young, this could have made youth more important than old age for early Christians, and early Christianity did little for the old, though the building of hospitals and asylums may have helped them. In the Middle Ages the young ruled the world; even the Popes were mainly young. There were exceptions: Charlemagne ruled until he was 72, and Enrico Dandolo, the twelfth-century Doge of Venice, is infamous for his role in the Fourth Crusade at the age of 90. But in 1380, when Charles V of France, died aged 42, he was regarded as already old.

Eastern civilisations in old times showed respect to the old. The high position of the old in China is due to Confucius (551–479 BC), who gave superiority to the elderly; for him the whole household owed obedience to the oldest man, who had the right of life and death over his children. On his 70th birthday Confucius said: 'I could follow the dictates of my heart without disobeying the moral law.' Confucius stated that filial piety ' . . . is the root of all virtue, and the stem out of which grows all moral teaching.' The Chinese wanted to grow old, or at least to appear old, because of the privileges enjoyed by older people. When two people of different ages were together, the elder spoke freely and the younger listened respectfully, so the younger man wished to grow older so that he might talk more and listen less. In family life, age brought authority. The young saw the advantage of honouring and obeying their parents; even the middle-aged could profit by the wisdom of the old and feel repaid for supporting them. Several Chinese proverbs illustrate this view: 'If you wish to succeed, consult three old people'; 'He who will not accept an old man's advice will some day be a beggar'; 'If a family has an old person in it, it possesses a jewel.' As a result, a man's 50th birthday in China was marked with reverence. Fathers had the right of life and death over their children.

The old are rarely seen in Western literature in the Middle Ages, though *La Morte D'Arthur* has the king aged over a hundred and Chaucer, in *The Canterbury Tales*, makes an old man's sexual activity revolting. Most seventeenth- and eighteenth-century writers are said to have viewed old age as a time of physical decline, and thought that old people were peevish, garrulous and forgetful. Daniel Defoe's *Moll Flanders* (1722) is one of the earliest novels in which a woman's life is followed until she reaches seventy, and which shows the strength of her character. Moll's lifestyle was lurid, but her closing words are that she and her husband had determined 'to spend the rest of our lives in sincere penitence for the wicked lives we have lived'. For Moll, age was a time of cheer and good humour, where one can make up for the failings of a lifetime. Goethe, who wrote *Faust* at the age of 65, also took a positive line:

> So, lively brisk old man
> Do not let sadness come over you;
> For all your white hairs
> You can still be a lover.

Charles Dickens objected to the comparision of old age to childhood, reckoning it as similar as death is to sleep. In *The Old Curiosity Shop*, little Nell's grandfather is very kind, but gambles too much. Victor Hugo's plays had many old characters with positive features. In Guy de Maupassant's short story 'All Over', an old man meets again a woman he loved, but is shocked by how she has aged – only her daughter resembles his early love.

Shangri La is a fictional place in James Hilton's 1933 novel *Lost Horizon*, which may have been inspired by Hilton's visit to the Hunza Valley in northern Pakistan, where the inhabitants have been reported to live long and healthy lives. Exercise is an important part of their life, as the mountains are extremely rough terrain. They eat mainly fruit and wheat, barley and millet. They have been called by some researchers 'The Happiest People on Earth'. The main characters in the novel are taken to a secluded monastery where the monks practice a combination of Christianity and Buddhism and where some are immortal.

The novel *Memento Mori* by Muriel Spark in 1958 marks the beginning of a sustained interest among novelists in what V. S. Pritchett called 'the great suppressed and censored subject of contemporary society, the one we do not care to face, which we regard as indecent: old age.' Earlier novels had often just included a mid-life decline. All the characters in *Memento Mori* are over 70, and most in their 80s, and are in a nursing home. The novel centres around the anonymous callers who phone and say, 'Remember that you must die.' Old age is presented as a confusing mix of feelings, memories and inabilities leading to death: 'I would be glad to be let die in peace. But the doctors would be horrified to hear me say it. They are so proud of their new drugs and new methods of treatment

– there is always something new. I sometimes fear, at the present rate of discovery, I shall never die.'

Since *Memento Mori* many novels have appeared with central figures over 70, but they deal almost exclusively with the isolation, impotence and decay which are regarded as intrinsic to the ageing process. The closing lines of the final story in John Updike's last book, *My Father's Tears*, describe a man in his late 70s raising the glass of water he uses to wash down his nightly medications – his cholesterol-lowering pill, the anti-inflammatory one, his sleeping pill, his calcium supplement – in a toast 'to the visible world, his impending disappearance from it be damned'.

W. B. Yeats was infuriated by old age, which he recognised as inescapable, but his famous poem is wonderful and encouraging:

> When you are old and grey and full of sleep,
> And nodding by the fire, take down this book,
> And slowly read, and dream of the soft look
> Your eyes had once, and of their shadows deep;
> How many loved your moments of glad grace,
> And loved your beauty with love false or true,
> But one man loved the pilgrim Soul in you,
> And loved the sorrows of your changing face;
> And bending down beside the glowing bars,
> Murmur, a little sadly, how Love fled
> And paced upon the mountains overhead
> And hid his face amid a crowd of stars.

T. S. Eliot's *Little Gidding* is pessimistic:

> Let me disclose the gifts reserved for age
> To set a crown upon your lifetime's effort.
> First, the cold friction of expiring sense

> Without enchantment, offering no promise
> But bitter tastelessness of bitter fruit
> As body and soul begin to fall asunder.

A recent survey by Age Concern found that older people are often stereotyped as 'warm and incompetent', or 'doddery but dear', and younger people are stereotyped as relatively cold but competent. The rating of young people as more competent than older people can perhaps be explained by attributing memory failure to laziness in the young but incompetence in the old. A key finding from the survey was that older people themselves hold self-stereotypes and values which are likely to result in age-based prejudice. Specific findings supporting this view are that people over 65 are as likely as the rest of the population to hold the 'warm but incompetent' stereotype of the old, and those over 75 particularly are more likely to agree that competence declines with age. Those over 75 are the least likely to want to extend equal opportunities for older people. Results overall showed that many people did identify with, and felt a strong sense of pleasure in belonging to, their own old age group, but about a quarter did not. Even the elderly population has a tendency to stereotype their own age group. While older people were stereotyped as friendlier, more admirable and more moral than younger people, younger people were viewed as more capable. In general people held more positive views about their own age group and almost all had most friends of their own age.

People across all age groups tend to agree that older people are admirable to some extent, and friendly to a greater extent. Moreover, older people see themselves as more likely to be viewed as moral, intelligent and capable than younger groups. They also see themselves as less likely to be viewed as pitiable or disgusting. A widely held view is that people over 70 should be valued and cherished; there is almost universal agreement on this. Many feel that equal employment opportunities for older people have not gone far enough. While American children have a positive view of older adults in their own family, they may have a negative world view of ageing. One explanation is that the stories the youngest of children are introduced to often portray older people as wicked or weird, like the evil old witch in 'Hansel and Gretel' and the scheming Rumpelstiltskin. Overall, children do have positive perceptions of the old. Older men are generally perceived more positively than women.

Only a quarter of those asked in a survey on work thought people over 70 were at all likely to be viewed as capable of working competently, compared with nearly half who thought people under 30 were likely or extremely likely to be capable. Other research findings, however, indicate that younger workers are often no better at their jobs than older workers, despite the widespread perception that this is the case. It has been shown in experiments that there is no significant difference between the abilities of younger and older workers, with each group performing particularly well or poorly in different areas. It is suggested that less good performance by the old due to reduced cognitive processing is counter-balanced by increased ability because of previous relevant experience.

Most people would be more comfortable with a suitably qualified manager of over 70 than one of under 30.

Almost half think that employers avoid having older people on their workforce because it spoils their image. It was generally accepted that a good way to reduce prejudice and discrimination between old and young groups is to foster close personal friendships between members of each age group. Good relationships between grandchildren and grandparents could certainly help.

Writing in the *Sunday Times*, the TV critic A. A. Gill offered a more forceful view, describing the old as 'zombies at the end of our own home horror movies . . . Ageing is so frightening in part because we treat the old so badly, and we treat them badly because we are so frightened of them . . . This is the greatest shame and horror of our society and of our age.'

Germans tend to view ageing much more negatively than Americans, and Americans consider themselves to be 'old' at a much younger age than Germans. Yet elderly people in the United States today are not treated with the respect and reverence to which they were accustomed earlier in history. The gerontologist David Hackett Fischer notes that literature from seventeenth- and eighteenth-century colonial America stressed deference and respect for the elderly. He maintains that the elderly were viewed with a feeling of deep respect and reverence, with contrasts with more modern views. Today the elderly have become virtual outcasts of society, many living on the fringe, often in retirement communities or in nursing homes.

In modern industrial societies emphasis and value are placed on youth, with advertising geared towards and glamorising the young. To the extent that advertising acknowledges the elderly individual at all, it attempts to make him or her appear younger. The elderly are victims of mistaken beliefs and irrational attitudes promoted largely through the various mass media. It has been claimed that the most flattering thing you can say to an older American is that he 'doesn't look his age' and 'doesn't act his age' – as if it were the most damning thing in the world to look old. But at least many of we oldies are looking very well, as we are repeatedly told.

Many negative but influential views about ageing continue to derive from the media, including films and TV as well as books. Simone de Beauvoir, in an important book on ageing, wrote:

> It is old age, rather than death, that is to be contrasted with life. Old age is life's parody, whereas death transforms life into a destiny: in a way it preserves it by giving it the absolute dimension. Death does away with time . . . I have never come across one single woman, either in life or in books, who has looked upon her own old age cheerfully.

She uses the example of Leon Trotsky to show that even the body's signals can be ambiguous, and there is a temptation to confuse some curable diseases with irreversible old age. Trotsky dreaded growing old and he was filled with anxiety when he remembered Turgenev's remark, one that Lenin often quoted: 'Do you know the worst of all vices? It is being over 55.' In 1933, when he was exactly 55 himself, he wrote a letter to his wife complaining of tiredness, lack of sleep, a failing memory; it seemed to him that his strength was going, and it worried him. 'Can this be age that has come for good, or is it no more than a temporary, though sudden, decline that I shall recover from? We shall see.' Sadly he called the past to mind: 'I have a painful longing for your old photograph, the picture that shows us both when we were so young.' He did get better, and he took up all his activities again.

John Updike, in *Self-Consciousness*, wrote:

> As I age, I feel my head to be full of holes where once there was electricity and matter, and I wonder if, when my head is all hole, I will feel any more pain or loss than I do now. What we don't know, we don't know: the Stoics are right at least about this. Ignorance is a kind of bliss, and senility, like drunkenness, bothers beholders more than the bearer.

In *Ageing and Society* (2000) Elizabeth Markson and Carol Taylor found 3,038 American films made between 1929 and 1995 which featured male actors over 60 years of age who had been nominated at least once for an Oscar. (We are used to seeing well-known older male actors wooing women stars half their age; it's less common – if not unknown – the other way round.) But a random sample of these films showed that whereas older men were portrayed as 'vigorous, employed and involved in same-gender friendships and adventure whether as hero or villain', women remained 'peripheral to the action or were portrayed as rich dowagers, wives/mothers or lonely spinsters'. They concluded that film roles have remained remarkably static in age and gender stereotyping despite changes in society.

Well-known films featuring older heroes include Ingmar Bergman's *Wild Strawberries*, which makes use of reminiscence to explore the disillusionment of an elderly physician as he reflects on his life and his mortality. *Driving Miss Daisy* won Jessica Tandy an Oscar at the age of 80 for her portrayal of a testy Southern Jewish woman's relationship with her chauffeur. *On Golden Pond* featured the veteran actors Henry Fonda and Katharine Hepburn as an elderly couple sorting out their relationship with an estranged child. More recently there have been films which have picked up on the theme of Alzheimer's, notably *Iris*, based on the life of the writer Iris Murdoch, and *Away From Her*, in which Julie Christie plays a sufferer who insists on entering a rest home, which greatly upsets her husband as he cannot communicate with her or visit her for a long period of time. And most unexpectedly, a 78-year-old man, a bit grumpy but tough and kind, is the hero of the 2009 Walt Disney animated film *Up*. He sets out to fulfil his lifelong dream to see the wilds of South America but he isn't alone on his journey, since an 8-year-old boy, a wilderness explorer who is trying to get a badge for assisting the elderly, has become a stowaway on the trip. They have amazing and amusing adventures, and encounter talking dogs, an evil villain and a rare bird. The boy gets his reward.

Research in the US has found that during prime-time television shows, only 3 per cent of the characters are aged 65 or older, while this age group actually accounts for 9 per cent of the American population. Older people portrayed on television are often marginalised, comical, or based on stereotypes. Fewer elderly women were shown, although the number of older women outnumbers that of older men. Television has featured the situation of older people in series such as the American *The Golden Girls*, which featured four older women sharing a home and earned multiple Emmy Awards, and the British series *As Time Goes By*, with Judi Dench and Geoffrey Palmer as a couple who meet again after a gap of 38 years. And of course there is the redoubtable figure of Agatha Christie's sharp-witted detective Miss Marple to remind us that not all old people need be peripheral to the action. In the long-running BBC radio serial *The Archers*, June Spencer, at the age of 90 still playing matriarch Peggy Woolley, was involved in a storyline about the dementia of her fictional husband which echoed her own experiences with her real-life husband; and Betty Driver was at 90 still paying a role in *Coronation Street*. The BBC sitcom *One Foot in the Grave* was so popular that its main character, Victor Meldrew, has become shorthand for a constantly bitter and complaining elderly man.

In politics, the standing of the old varies widely between different societies. Governments based on rule by the elderly – gerontocracy – have been common in Communist states, in which the length of one's service to the Party was held to be the main qualification for leadership. In the time of the Eight Immortals of the Communist Party of China, who held much power in the 1980s, it was quipped that 'the 80-year-olds are calling meetings of 70-year-olds to decide which 60-year-olds should retire'. For instance, Party leader Mao Zedong was 82 when he died, while Deng Xiaoping retained a powerful influence until he was nearly 90. In the Soviet Union, gerontocracy became increasingly entrenched from the1970s, at least until March 1985, when a young, ambitious government headed by Mikhail Gorbachev took power.

The public may not always be keen on old politicians. Sir Menzies Campbell was 64 when he was elected leader of the Liberal Democrats in 2006. Cartoons in the newspapers made him old, bald, derelict, and looking 150.

The media went for his age, which made him, it was claimed, unacceptable and not suitable for the job; the *Financial Times* said leaders had to be young. He was repeatedly asked whether he was just too old for the job. He vigorously defended the advantages of aged people and argued that their experience was very valuable, but he was forced out of office. Similarly, at 72 John McCain was regarded by many as too old to be the next US president – far too long in the tooth. Had McCain succeeded in his 2008 campaign he would, at 73, have been the oldest President in American history. Discussions about his age dogged McCain during his failed run, and people recalled that Ronald Reagan showed early signs of Alzheimer's in his late 70s. Yet the president of Egypt, Hosni Mubarak, at the age of 81, has been in power for 28 years. The economist J. K. Galbraith, at 87, was irritated by ageist remarks like a 'Are you still working?' And 'Are you taking exercise?' To those who asked such questions he wanted to reply with 'I see that you are still rather immature.'

The young, not the old, benefited in the 60s from post-war affluence in the West. Youth began to develop its own culture and the young of the 1960s did not want to lose the benefits. Cosmetic sales to hide ageing in the USA went up some tenfold in this period. Fitness became popular and women began to refuse to accept their old-age stereotype. Advertising focused attention on the third age, and there were magazines directed to older customers, but the old were dismissed from most of public life. Roger Daltrey in the 1960s sang 'I want to die before I get old', and Timothy Leary advised those on the campus to ignore anyone over 30. Many of us, when looking at the old when we were young, did not believe that it would happen to us.

One attempt to produce an antidote to youth culture is *The Oldie*, a monthly magazine launched in 1992 by Richard Ingrams, who for 23 years was the editor of *Private Eye*. It carries general interest articles, humour and cartoons and is sometimes regarded as a haven for 'grumpy old men and women' – an image it has played up to over the years with such slogans as '*The Oldie*: Buy it before you snuff it' and its lampooning of 'yoof culture' and the absurdities of modern life.

It is encouraging for those who fear ageing that nearly half of Americans aged 65 and older, when questioned, described the present as 'the best years of my life'. But at the same time many of the comments made by the elderly about themselves do not stray far from the stereotypes: 'My body's ugly obstinacy in keeping on living strikes me as admirable'; 'What do I want? Money and a younger woman'; 'In myself I observe the very traits that used to irritate me in men of late middle age whom I have known: a forgetfulness, a repetitiveness, a fussiness with parcels and strings, a doddery deliberation of movement with patches of inattention . . . I feel also an innocent self-absorption, a ruminativeness that makes me blind and deaf and indifferent to the contemporary trends and fads that are so crucial to the young'; 'Every attempt to be specific about the afterlife, to conceive of it in even the most general detail, appalls us.'

Not atypical current attitudes to getting old come from a recent article by Tim Lott, who is in his 50s, in the London *Evening Standard*. He points out that the novelist Kazuo Ishiguro, now 65, thinks writers are over the hill when past their 30s, and that a woman of 34 feels she is already old. Lott recognises that many great writers flowered in later life but says:

> There are disadvantages to growing old – you smell, your teeth crumble and the bad habits that you once thought you could rid of by sheer force of willpower you now realise are as inescapable as your rumpled skin. But the great consolation is that all your contemporaries are crumbling in much the same way. Even people who were once rock stars can now be joyfully observed on TV resembling balding retired pork-pie tasters.

In another piece, he says the advantage of being older is that at last you know who you are. If you are then ugly, so are all your contemporaries. You also probably have more money.

The charity WRVS, which supports volunteers working for the elderly, found that 40 per cent of the public don't feel that they do enough to support older people, 65 per cent of people feel that older people make a positive contribution to society, and 76 per cent of people feel that older people are not treated with respect; they claim that the elderly are perceived as unhelpful and rude. 'Silly old goats growing old disgracefully' is how the elderly were labelled in a newspaper article describing their activities at a party. Another similar recent anti-age remark can be found in the list of the world's most livable cities: Vienna comes top of the list, but a negative comment is that it is full of grumpy old fur-coated ladies.

'You cannot teach an old dog new tricks' is but one of many proverbs about the old. In spite of the numerous tales and proverbs celebrating the wisdom of old people and promoting their care, folklore is replete with reflections of a basic distrust of age. The fear of the old is further reflected in the fairy tales of many countries in which old women, even those who at first appear to be helpful and kindly, frequently turn out to be sinister witches. Various demonic personages, notably changelings and the devil himself, can be rendered powerless by tricking them into revealing their age. Parents cannot necessarily expect the same care in their old age that they earlier tendered to their children. As the proverb has it 'One father can better nourish ten children than ten children can nourish one father.'

In an Irish folktale a man has a father who has grown too old to do anything but eat and smoke, so the man decides to send him away with nothing but a blanket. 'Just give him half a blanket,' says the man's son from his cradle, 'then I'll have half to give you when you grow old and I send you away.' Upon hearing this, the man quickly reconsiders and allows his old father to remain after all, saying: 'Good deeds are wasted on old men and on rogues.' Another man in the prime of life abuses his ageing father; he strikes him and drags him out of the house by his hair. When he too becomes old his son treats him the same way. One day the son drags him out the door and on to the street. 'You go too far!' cries the old man. 'I never dragged my old father beyond the gate.'

Many attitudes towards the old are deeply ingrained, recurring from one generation to the next. How they effect the practical ways in which the old are treated and cared for will be discussed next.

11
Mistreating

'Ageism is as odious as racism and sexism' – Claude Pepper

Herr Levin von Schulenburg, a high official in Altmark, was travelling in about 1580 when he saw an old man being led away by several people. 'Where are you going with the old man?' he asked, and received the answer, 'To God!' They were going to sacrifice him because he was no longer able to earn his own living. When the official grasped what was happening, he forced them to turn the old man over to him. He took him home with him and hired him as a gatekeeper, a position that he held for 20 additional years.

Geronticide – the killing of the old when they are no longer of any use – features in the folk tales of many lands but has also been a historical reality. Even today some cultures do not encourage the survival of the old, much less suffer their continued burden. It is exceptional that in some primitive tribes the old are revered and cherished. The toughness of life, and scarcity of food, can render hearts impervious to soft sentiments with respect to the old. Particularly nasty examples have included claims of the killing of the old in several indigenous societies such as the Inuits, who live in the Arctic, the last example being in 1939. It is not clear how reliable these reports are.

The term 'ageism' was introduced in 1969 to refer to a combination of prejudicial attitudes towards older people, the promoting of negative stereotypes of old age, and discriminatory practices against older people. But as we have seen, it has a very long history. One of the comments about prejudice against the old before the term was in common use was by Max Lerner in 1957: 'It is natural for the culture to treat the old like the fag end of what was once good material.' The psychologist Dominic Abrams has claimed 'Ageism is the most pervasive form of prejudice experienced in the UK population and that seems to be true pretty much across gender, ethnicity and religion – people of all types experience it.' Ageism was described in 1975 by Robert Butler 'as a process of systematic stereotyping of, and discriminating against, people just because they are old'; and by R. C. Atchley, as 'a dislike of ageing and older people based on the belief that ageing makes people unattractive, unintelligent, asexual, unemployable, and senile'. He claims that research indicates that most Americans subscribe to at least a mild form of ageism.

A major example of ageism and age discrimination in everyday life in the UK is the mandatory retirement age set at, or after, the age of 65, though the mandatory retirement age for civil servants has been abolished. Early retirement is not necessarily a good thing for an individual. Over 100,000 people were recently forced to retire against their will and this has made life very difficult for many of them. With an estimated 120,000 older workers forced to retire in 2009, this policy is draining billions of pounds from the economy every year. Forcing over 100,000 employees out of the job market has opened up an estimated £3.5 billion gap in lost economic output, inclusive of £2 billion in lost earnings for the workers themselves. But the government has pledged to get rid of the mandatory retirement age.

A survey in the *Economist* of articles involving ageing over a recent 10-year period found that most showed a predominantly ageist view of older people as a burden on society, often portraying them as frail non-contributors. Costs of healthcare for old peope are regularly viewed unsustainable and pensions as a demographic 'time bombs'. Over time small increases in average life expectancy can lead to very large increases in the size of a population, but have also resulted in large gains in economic welfare over the past century; these gains are consequences of improvements in life and health expectancy and are not restricted to a handful of old people.

Work in later life can contribute to older people's health and wellbeing and can make a dramatic difference financially: ten more years of working life can double the value of a typical private pension. Magistrates and jurors are not allowed to serve past the age of 70, and older workers are rated consistently lower than younger workers, despite no significant differences in work achievements. In fact older workers are actually more reliable in terms of absenteeism than younger workers. The Employment Equality Act (Age) Regulations 2006 has made it unlawful to discriminate in a work or training context against someone because of their age. A worker should not be disadvantaged in any area of employment such as recruitment, employment benefits and dismissal. But there are still crucial exceptions, the main one being that this does not apply to those over 65.

If a task genuinely has to be done by someone who has a particular characteristic related to age, it is currently lawful to discriminate in order to achieve this. An example would be the case of an actor having to play a role of a young character – here there can be discrimination against old actors and a young one can be selected. But it is unlawful for an employer to discriminate against someone because he or she is over 50 unless they can justify their actions, or it is covered by one of the exemptions included in the law. An example of discrimination could be the case of a job applicant aged 60 who has evidence to show that she is better qualified than the person who got the job, who is aged 35, even though the job advert stated that the employer was looking for a 'junior manager'. The application form had asked for her date of birth. Pay and benefits should be based on skills, and not age.

In the USA the Age Discrimination in Employment Act of 1967 protects individuals who are 40 years of age or older from employment discrimination based on age. It permits employers to favour older workers based on age even when doing so adversely affects a younger worker. An article in the *Wall Street Journal* claimed that as unemployment intensifies in the economic downturn, claims of age discrimination are soaring. Although mandatory retirement has been abolished in the USA, there are certain types of jobs that do have mandatory retirement laws. These are jobs that are too dangerous for older people or jobs that require particular physical and mental skills. Some of these jobs are those of military personnel, fire fighters, airline pilots and police officers. But retirement is not based on an actual physical evaluation of the person, and this is why many people consider mandatory retirement laws for these jobs to be a form of age discrimination.

Ageism is more than simply having negative attitudes about old age. It can include the following: being refused interest-free credit, a new credit card or car insurance because of age; an organisation's attitude to older people resulting in them receiving a lower quality of service; age limits on benefits such as Disability Living Allowance; a doctor deciding not to refer an old patient to a consultant; losing a job. Age discrimination can involve serious negative treatment that can affect how one lives.

A major biennial survey of over 2,000 adults run by Age Concern since 2004 exposes the full extent of age discrimination in the UK. It reveals that more than three times more people have been the victims of ageism than any other form of discrimination. Direct discrimination occurs when an employer treats a worker less favourably than other workers on the grounds of age. Three fifths of people aged 65 and over believe older people suffer widespread age discrimination, including in the workplace.

Not surprisingly there was strong support for the Labour minister Harriet Harman's attack on ageism in a new Equalities Bill which has not been made law, but which advocated ban on age discrimination in provision of goods, facilities, services and public functions. Such a measure could be a milestone in the battle for fairness in later life. But as Age Concern point out,

> The Bill only gives ministers the power to ban age discrimination in services if they wish. We want to see an unbreakable legal commitment to introduce new rights, across the public and private sectors. Age discrimination in health and social care services can literally mean the difference between life and death. Because

> of their age, older people are being denied vital treatments with no legal protection. Each day older people are refused financial products like travel insurance for no better reason than the date on their birth certificate.

Research indicates that most Americans subscribe to at least a mild form of ageism. International Longevity Center-USA found that a majority of older adults reported that they'd been ignored or experienced insensitivity, impatience and condescension from others based solely on their age. The outcome can be more than just an embarrassing situation. Research shows that individuals receiving such treatment often end up with debilitating lowered self-esteem and self-confidence, as well as substandard healthcare. While nearly a third of those in the survey had experienced ageism in the last year, those over 65, and particularly those over 75, were less likely than the rest of the population to view age discrimination as serious.

The former deputy prime minister Lord Heseltine, aged 76, claimed that Britain is becoming an ageist society 'worshipping at the altar of the young', and blamed a 24-hour news culture for perpetuating stereotypes about young and old. He said:

> To my mind, 'old' is first and foremost something mature, ripened and proven, something that has survived time's test. But I appreciate that society as a whole, and the media especially, does not see it that way.

Rabbi Julia Neuberger believes that in the UK many older people are demeaned and made to feel worthless. It has been suggested that ageism is worse than racism or sexism because there is so little recognition that it is wrong. Discrimination on the basis of age is under-researched compared with racism or sexism; the most serious form of prejudice is considered to be racism, followed by prejudice based on disability.

Abuse of the elderly is the most serious problem of ageism. As many as half a million elderly people in the UK may be being abused, according to a House of Commons report in 2004. It found two thirds of the cases of abuse occur in people's own homes, and take the form of sexual, physical and financial abuse, neglect and over-medication. Much abuse is not reported because many older people are unable, frightened or embarrassed to report its presence. Often care staff take no action because they lack training in identifying abuse, or are ignorant of the reporting procedures.

A UK Study of Abuse and Neglect of Older People in 2006 found that in the past year about 227,000 people aged 66 and over living in private households reported that they had experienced mistreatment involving a family member, close friend or care worker. Mistreatment by neighbours and acquaintances was reported in about one third of cases. Overall, half of mistreatment involved a partner or another family member. About 10 per cent involved a care worker, and 5 per cent a close friend. Most of those responsible for physical, psychological and sexual abuse were men, while financial abuse was spread more equally between the sexes. Three quarters of those asked said that the effect of the mistreatment was either serious or very serious, and left the person feeling upset and isolated. About one third told nobody but most told family, friends, or a social worker or health professional. Very few informed the local authority or the police.

In developing countries there is no systematic collection of statistics about abuse, but crime records, journalistic reports, social welfare records and small-scale studies contain evidence that abuse, neglect, and financial exploitation of elders are much more common than these societies admit. A hospital manager in Kenya was quoted as saying: 'Older people are a big headache and a waste of resources. The biggest favour you could do as an older people's organisation is to get them out of my hospital.'

In healthcare, ageism presents serious problems as the National Service Framework for Mental Health applies only to people below the age of 65. This seems to be a clear case of age discrimination, particularly as dementia affects large numbers of people over 65. Those over that age receive lower-cost and inferior services to younger people, even if they have the same condition. Patients older than 65 are being denied treatments offered to younger people, either because they are too expensive or because they were not referred on by their family doctor.

There is considerable evidence of discrimination against the elderly in healthcare, with staff disbelieving older people's accounts of their medical or clinical symptoms, or with these being disregarded as a natural condition of their age. Older hospital patients can be seen as financial risks: they are viewed less as human beings with health needs than as costly and inanimate 'bed-blockers'. Physicians themselves and other healthcare providers may hold attitudes, beliefs and behaviours that are associated with ageism against older patients. Studies have found that physicians often do not seem to show concern in treating the medical problems of older people. There is evidence that more than a third of physicians erroneously consider high blood pressure to be a normal part of ageing, and do not treat the condition in their older patients.

The average size of care home for older people in England is 34, compared to nine places in homes for younger adults. Many of the old themselves believe that doctors view them less favourably than younger patients. Fewer than 10 per cent of older people with clinical depression are referred to specialist mental health services compared with about 50 per cent of younger adults with mental and emotional problems. Elderly stroke patients treated in the NHS do not get the same level of care as younger patients, who are scanned more quickly and more often. Mental health wards for older patients are less clean, more noisy and more violent than average.

Rabbi Julia Neuberger and many others have argued for legislation against such age discrimination. Older patients are less likely to have their symptoms fully investigated. A study by the Patients Association reported that some NHS nurses had been shockingly cruel to the elderly; some had been left without food or drink while others had been made to sleep in soiled bedclothes. It was estimated that there had been up to a million such incidents in recent years. Though there may be surgeries or operations with high survival rates that might cure their condition, older patients are less likely than younger patients to receive all the necessary treatments. It has been suggested that this is because doctors fear their older patients are not physically strong enough to tolerate the curative treatments and are more likely to have complications during surgery that may end in mortality. The approach to the treatment of older people is often concentrated on managing the disease, rather than preventing or curing it. Thousands are discharged from hospital too early.

Some sources suggest that ageism in the healthcare system starts in the medical schools where young people – who, of course, will never themselves be old – begin their education. Only 10 per cent of medical schools in the United States require courses in geriatrics and less than 3 per cent of physicians ever take any courses in this area. In the UK there are some medical schools that do not teach geriatric medicine. When actually interacting with older patients on the job, doctors sometimes view them with disgust and describe them in negative ways, such as 'depressing' or 'crazy'. For screening procedures, elderly people are a bit less likely than younger people to be screened for cancers and so less likely to be diagnosed at early stages of their conditions.

Outside the healthcare system, Help the Aged reported that older people routinely tell them that they feel ignored and undervalued by their local communities. There needs to be much more energy and determination to reach those who are seldom heard, for example isolated older people and those living in poverty. By contrast, fortunately, many very old people are sustained by love and care of family and friends.

Some forms of ageism are described as 'benevolent prejudice' because the tendency to pity is linked to seeing older people as 'friendly' but 'incompetent'. This is similar to the prejudice most often directed against women and disabled people. Age Concern's survey revealed strong evidence of 'benevolent prejudice'. The warmth felt towards older people means there is often public acceptance that they are deserving of preferential treatment – for example, concessionary travel. But the perception of incompetence means older people can also be seen as 'not up to the job' or 'a menace on the roads' when there is no evidence to support this. Benevolent prejudice also leads to assumptions that it is 'natural' for older people to have lower expectations, reduced choice and control, and less account taken of their views.

When older people forget someone's name, they are viewed as senile, but when a younger person fails to recall a name, we usually call that a faulty memory. A newspaper recently reported that the actress Keira Knightley was having something akin to a 'senior moment' as she came off a plane from London, due to the seven-hour flight and a five-hour time difference. She had just forgotten something, but this phrase is a mild example of ageism. Use of the term implies that she was suffering from one of the problems that afflict the elderly. I am all too well aware of them. But the young also forget things.

When an older person complains about life or a particular incident, they are called cranky and difficult, while a younger person may just be seen as being critical. It is quite widely assumed that older people might not want the sorts of life chances that younger people have and so it is 'natural' for older people to have lower expectations, and less account is taken of their views. Fortunately, as we have also seen, older people are further stereotyped as moral and admirable, and an overwhelming majority of people agree that they should be valued and cherished. But although most people think older people should definitely have equal access to health and care, this is often not the reality.

One way that implicit or explicit ageism may manifest itself is through the use of patronising language with older people. There are many ageist articles in the media, often presenting the elderly as a burden on younger people in families and society at large. Critical analyses have suggested that both negative and positive newspaper portrayals of old people may be ageist.

Negative stereotypes of older people range from the hostile image of a 'cantankerous old codger' to less explicit images. Not uncommon ageist terms inlude 'old fogey', 'old fart', 'geezer', and 'old goat'; even the word 'old' itself is often used as an insult. Elderspeak refers to a way of communicating with the elderly – it is simplified language with exaggerated pitch and intonation. This can be based on beliefs about the elderly and personal experience.

The term 'patronising language' specifically describes two negative methods of communication: the person being unnecessarily courteous and speaking simple and short sentences loudly and slowly to an older person, with an exaggerated tone and high pitch; and baby talk, which involves the exaggerated pitch and tone that one uses when talking to a baby. Both these ways of talking have negative effects on the elderly. Anti-ageism activists in the US have strongly argued against the use by journalists of terms such as 'elderly', 'fogey' or 'codger' – and even 'senior'. They recommend the avoidance of phrases such as 'of a certain age', and 'old ladies' of both the 'little' and 'sweet' variety. The advice is included in a media guide on reporting issued by the International Longevity Center and Aging Services of California.

In this guide, the campaign attempts to help journalists and advertisers represent 'older people' – its preferred term – in a 'fair contemporary and unbiased' manner. The authors state that 80 per cent of older Americans have been subjected to ageist stereotypes. While names and characterisations may vary, the message is the same: older men and women are incompetent and lack sufficiency. Journalists are advised: 'If you need to identify individuals over the age of 50, "older adults" is preferred over "senior" and "elderly", which can be discriminatory in nature as we do not refer to people under 50 as "junior citizens". If relevant to the story, state the age.' Out goes 'golden years' as a description of an individual's period of life after being deemed to be an older adult.

Images as well as words may be ageist. Campaigns have been mounted in Denmark and some other countries to counter images of old people shown as overweight or sickly. In Australia money was given to promote the contribution of the old to social life: 'Look past the wrinkles' was on a billboard in Melbourne.

The stereotypes and infantilisation of older people by patronising language affects older people's self-esteem and behaviour. Ageism, as distinct from discrimination, has significant effects. Exposure to ageist stereotypes has negative affects on physiology and mental abilities. After repeatedly hearing that older people are useless, older people may start to perceive themselves in the same way that others do, as dependent, non-contributing members of society. Studies have specifically shown that when older people hear about their supposed incompetence and uselessness, they perform worse on measures of competence and memory. These negative stereotypes thus become self-fulfilling prophecies. Then this behaviour in turn reinforces the present stereotypes and treatment of the elderly. Negative attitudes towards older adults and stereotypes about older people emerge early in a child's life, even in such a simple ways as, for example, selecting a younger adult to partner them in a game rather than an older adult.

Ageism operates in high-profile professions no less than in others. In a recent interview, actor Pierce Brosnan, aged 57, cited ageism as one of the contributing factors as to why he was not asked to continue his role as James Bond in the Bond film *Casino Royale*, released in 2006. Successful singer and actress Madonna spoke out in her 50s about ageism and her fight to defy the norms of society. The actress Geena Davis, at 52, complained that she could not get a decent role because of her decrepitude. Joan Bakewell, who was appointed by the government as the 'voice of older people', has criticised the BBC for banishing female news presenters once they reach 50, and for the general lack of older female faces. Arlene Phillips was dropped from *Strictly Come Dancing* when she was 66. The journalist John Simpson has also complained about ageism in the BBC. Yet just how impressive older women can be was shown when the 89-year-old novelist Baroness P. D. James interviewed the director of the BBC on the *Today* programme about his executives' pay packages and, according to press reports, reduced him to a stuttering wreck. But a very recent tribunal ruled against the BBC for dismissing a 53-year-old female presenter.

In the US version of *The Weakest Link*, contestants' voting decisions were, on average, biased against older panellists. At the stage of the game where it is in participants' interests to vote for poor performers, older people were likely to be chosen even when younger adults had performed worse. But when contestants would benefit by choosing top-performing rivals to eliminate them from the competition, they tended to choose lower-performing, older contestants. Subconsciously, the panellists simply did not want to be around older people.

In spite of its negative effect on the daily lives of older people, ageism is often unrecognised, ignored or even compounded in health and social care settings. And social exclusion has only recently been officially acknowledged as affecting older people as well as children and families.

Older consumers have grown into a market force to be reckoned with, says Age UK, as new figures reveal the amount of money spent annually by people over the age of 65 in the UK is set to hit the £100 billion mark. Yet despite seeing their weight in the consumer market grow as a group, older people are still at risk of being frozen out of a marketplace which is slow to adjust to the evolution of an ageing society. Research by the charities found that many older people think businesses and retailers have little interest in the consumer needs of older age groups, and many still face obstacles in accessing financial services which are tailored to the needs of younger customers.

There are other, more robust views. The actress Joanna Lumley proclaimed, 'I am not being unkind but I am just saying millions of crones like me shouldn't suddenly be given the lead in things just because we are damn old.' But this laudable refusal to claim special privileges for the old should not blind us to the rights of the old to be treated fairly, or to the fact that ageism can lead to social exclusion, diminish the quality of life which older people may enjoy, and threaten their mental health.

12

Caring

'Old age ain't no place for sissies' – H. L. Mencken

Joan Bakewell put the problem of care very simply:

> It is what we all hope for. To enjoy our final years living in our own homes, among familiar things, not only the worldly goods but also the memories that cling to the place we know. We hope that when our time comes, some system will be in place that brings sympathetic and well-trained people to our homes, visiting at agreed times, bringing care and comfort, and whatever medical attention we will need. We probably don't mind too much who provides that care as long as it is friendly and professional. Not much to ask?

Alas, it seems too much to get. A rather negative view came from Michelle Mitchell, when a director for Age Concern:

> Loneliness, depression, poverty and neglect blight the lives of millions of older people and for many, evidence shows the situation is getting worse, not better. Attitudes to older people are stuck in the past, the care and support system for older people is on the brink of collapse and older people's experiences of isolation and exclusion have largely been ignored by successive governments.

Research conducted by the Joseph Rowntree Foundation investigated what today's older people consider to be necessary for 'comfortable, healthy ageing'. The major conclusion is that courage is required to cope with the ageing process, either in fighting it or adapting to the limitations it brings.

There is no common way in which societies care for their old today, and attitudes have changed with time, but it used to be common for children to look after their ageing parents. In Ancient Greece, though there were negative attitudes towards the old, it was the sacred duty of the children to look after their parents or grandparents, and Greek law laid down severe penalties for those who failed to discharge their obligations. In Ancient Rome, the old had great privileges within the family; both a father's and a grandfather's consent were necessary for a son to marry.

Shakespeare's *King Lear* gives a moving account of the lack of care for an old parent. Lear had visualised a world in which old men would continue to be respected even after giving away their money and their power. His two elder daughters, instead of looking after him, do all they can to alienate him and send him out, homeless, to the heath. 'Thou shouldst not have been old till thou hadst been wise,' Lear's Fool tells him. Should sons and daughters be responsible for their mothers, fathers and grandparents? This play's special understanding of old age partly accounts for it being among the most moving of Shakespeare's tragedies.

A short tale by Tolstoy is relevant to the predicament of Lear:

> A raven was carrying his chicks, one at a time, from an island to the mainland. In mid-flight he asked the first, 'Who will carry me when I am old and can no longer fly?' 'I will,' answered the

young raven, but the father did not believe him, and dropped him into the sea. The same question was put to the second chick. He too replied, 'I will carry you when you are old,' and the father also let him fall into the sea. The last chick received the same question, but he answered, 'Father, you will have to fend for yourself when you are old, because by then I will have my own family to care for.' 'You speak the truth,' said the father raven, and carried the chick to safety.

People of 75 have had a fair share of life, and many do very well looking after themselves with pleasure until much older. Nearly three quarters of those over 65 in the UK are home-owners. Many are well cared for in their own homes and care homes, and well looked after by the NHS, but by no means all of them. Some 40 per cent of those in care homes have been reported to be depressed. About two thirds of hospital beds in the UK are occupied by the over-65s in all the different wards, and dementia and depression are the most common ailments. Falls and incontinence are also serious problems. Clearly this is very costly, and 40 per cent of the NHS budget is spent on age-related illnesses.

Ageing makes self-support increasingly difficult. About one in four of the elderly in developed counties will need long-term care of some kind. How much do those who do need help deserve? The Japanese ex-prime minister Taro Aso lost a lot of support when he questioned whether it was right to put large sums of money into healthcare for the elderly. He is reported to have said: 'Why should I pay tax for people who just sit around and do nothing but eat and lounge about drinking?' In the UK more than one million older people get some local-authority-supported community-based care.

The concept of care and comfort for the old who need help has a long history: 'And he shall be unto thee a restorer of thy life, and a nourisher of thine old age: for thy daughter in law, which loveth thee, which is better to thee than seven sons, hath born him' (Ruth 4:15). There is also a plea for shelter for the aged in the Talmud, and by the eleventh century, these exhortations had led to the development of Jewish Homes in France and Germany to house the aged. Even before this, during the Byzantine Empire (AD 324–1453) the care of older persons had been undertaken in welfare institutions.

In England the first recorded almshouse was founded by King Athelstan, the first king of all England, and Alfred the Great's grandson, in York in the tenth century. Almshouses are charitable housing provided to enable people, typically elderly, who can no longer work to earn enough to pay rent, to live in a particular community. The Poor Law Act of 1601 provided some relief for those too ill or old to work, the so called 'impotent poor', in the form of a payment or items of food or clothing. Some aged people might be accommodated in parish almshouses, though these were usually private charitable institutions. The policy was to make the old and infirm as comfortable as they could be, and the able-bodied, if they managed to get in, uncomfortable. The Poor Law also created workhouses whose inmates had to go out and work. Then the New Poor Law of 1834 was enacted to make life harsher for those living in workhouses, so that they would prefer to be elsewhere. In 1871 payments for relief were reduced and children were expected to support their old parents. Some 2,600 almshouses continue to be operated in the UK, providing 30,000 dwellings for 36,000 people.

In France, the concept of convalescent homes was developed with the Hôtel-Dieu (originally founded in 651) and the Hôpital de la Charité in the seventeeth century. The first nursing homes in the United States were charitable institutions run by Catholics or Jews, in 1842. In 1853 Charless House, a charitable institution, was opened as a home for the friendless in St Louis. While intended to look after women of all ages, the persons admitted were predominantly older widows.

In the late nineteenth century old age began to be viewed as an illness. Being in a workhouse was degrading and many preferred death. Emmeline Pankhurst, a leading suffragette, described her experiences as a Poor Law Guardian in her autobiography *My Own Story*: 'I found the old folks in the workhouse sitting on backless forms, or benches. They had no privacy, no possessions, not even a locker. After I took office I gave the old people comfortable Windsor chairs to sit in, and in a number of ways we managed to make their existence more endurable.' In 1947 a Nuffield Committee argued that the character of workhouses needed to change, and that elderly persons should be accommodated in small homes to enhance their care.

The Second World War made life much harder for older people. As families became separated, or lost their main breadwinner, the problems faced by older people were compounded. People soon realised that the 'poor law' provision of the time was woefully inadequate. In 1940 a group of individuals, as well as governmental and voluntary organisations, came together to discuss how this situation could be improved and formed the Old People's Welfare Committee. With the birth of the welfare state in the 1950s, local and national government money became available to fund local work with older people. In 1971 the committee became completely independent of government and got a new name – Age Concern. High unemployment in the early 1980s caused Age Concern to join in government job-creation and training schemes. It drew attention to the plight of older workers who were unable to return to work because of long-term unemployment or redundancy. Age UK, a new charity combining Age Concern and Help the Aged, came into being in spring 2010.

The transition of the term 'elders' to 'the elderly' was probably due to the industrial revolution and improved health, leading to an increase in the number of older people. The economy had to be revised to accommodate an increased share of the population no longer in the work force. The first state pension in Britain was paid on the 1 January 1909, recognising the needs of the elderly. But it was just 5 shillings a week – £19.30 in today's money – and went only to the poorest half a million aged over 70. You were only eligible for the new payment if your income was less than 12 shillings a week, and the pension could be reduced if you had too much furniture.

In the present system calculating your state pension is quite complex. The state pension age for men is 65, and that for women is increasing from 60 to 65 in 2020. For both men and women, the state pension age will then increase from 65 to 68 between 2024 and 2046. A single person on a basic state pension currently gets a maximum of £97.65 a week. The Over 80 Pension is a state pension for people aged 80 or over who have little or no state pension.

Before the Second World War there was virtually no interest in old peoples' mental or physical health. Old age psychiatry was recognised as a speciality by the Department of Health only in 1989, but it is now a rapidly growing speciality within psychiatry. Dementia often brings critical needs for care, but sufferers who own more than £23,250 in property and savings will find that they must finance most of their care themselves. About half of all hospital and community care spending in England is for those aged 65 and over. Most of the old prefer to remain in their own homes and this often requires support from a carer if there is no support from the family. The preference of the elderly to remain at home is almost universal, but circumstances can make it necessary to leave. There is increasing diversity in family structure – in the UK about one third of the elderly have no children who could help caring for them. Divorce and single-parent families have also led to a decline in the traditional family. The relationship between ageing parents and their children and close relatives is complex. Allowing an elderly parent to live with the family may take up both valuable space and import some serious interference. Looking after a parent with dementia can be a nightmare. One understands the viewpoint of Agatha Christie: 'I married an archaeologist because the older I grow, the more he appreciates me.'

Even so, there is a very high contribution by families to caring for the elderly. An estimated 6 million people in Britain provide unpaid care for elderly spouses, parents or disabled children. That's nearly 10 per cent of the population. These carers get virtually no financial reward, though there have been unfulfilled promises by the previous government. They rarely get even one week's holiday. Huge numbers of carers make a contribution to caring in later life. Wives and daughters are more likely to help with caring, as males hire carers rather than doing it themselves. Over 1.5 million carers are themselves aged over 65 and an estimated 8,000 are over 90. Increasingly, care involves very frail couples, where the traditional boundary between carer and cared-for becomes blurred. At present individuals and the state split the costs of paid care roughly 50/50. But paid-for care is dwarfed in scale by support provided by unpaid carers, usually family members. In April 2010 the Labour government published a White Paper on care for the elderly that proposed that no pensioner will have to stump up the fees for care homes if they are there for more than two years.

As many as one third of the over 85s need help climbing steps, and a quarter with bathing or showering. Both the government and local authorities want many more people to stay in their homes with support, rather than going into more expensive care homes. That policy has been welcomed by campaign groups, but they are now questioning whether enough effort is going into regulating home care, and whether local authorities are trying to provide it on the cheap.

There are over 21,500 care homes, nursing homes and residential homes providing adult and elderly care throughout the UK. In England an estimated half a million physically disabled elderly are living in care homes or long-stay hospitals, and most of these are paid for by the state. The sick are treated free on NHS, so the NHS is responsible for meeting the full cost of care in a care home for residents whose 'primary need' for being in care is health based. But if one is simply frail and needs to enter a care home one must pay, unless one's assets are less than £23,250. A place in a care home in England costs an average of £24,000 per year while a nursing-home place costs an average of £35,000. Placing a relative in an average nursing home costs more than sending a child to Eton, one of the most expensive public schools in the land. It is estimated that older people spend in total about £6 billion for care out of their own money, and that the net spend by public authorities is similar.

Some one in four of the elderly in developed countries will need long-term care of some kind, an enormous number. In the US, most elderly Americans live in homes they own – 90 per cent live in their own home and most are satisfied with their living arrangements, and nearly half of those over 85 still live in their own homes. Just 4 per cent of adults aged between 75 and 84 live in an assisted living facility, and for those aged 85 and above it is 15 per cent. About half of retirees in the US expect to not move from their home. For decades, Americans have depended on nursing homes to care for them in old age. But as the population rapidly ages, more care is shifting from institutions to homes, and more responsibility is shifting to families – a change of major proportions. Sweden has one of the oldest populations – 5 per cent are 80 or older, and, remarkably, 94 per cent of all Swedes over 65 years of age still live in flats or houses. When planning housing and housing areas, Swedish municipalities are required to ensure that they are adapted to the needs of older people and those with disabilities. A further goal is for commercial and public services to be easily accessible so that the elderly can continue living at home and looking after themselves.

In China, it is seen as a great shame to put a parent into a nursing home and many are living alone, since there is only one child per family to look after them. The cost of sharing a home with a relative is, for the most part, borne by them and not the state. There are some advantages, such as grandparents caring for the young while the parents are at work. China has more than 40,000 elderly care institutions with about 1.7 million beds for its population of 145 million over 60, of whom over two thirds live in rural areas. Dementia is very high in China and so many more nursing homes are needed. About 13 million people aged over 80 are now in dire need of care. About one third or half of the elderly in large cities are without support from their children or a livelihood.

For Holocaust survivors, ageing presents an enormous challenge. In Israel there are 50,000 Holocaust survivors living below the poverty line. Having survived Hitler and the Nazis, they are now struggling with a new obstacle, the ageing process. Many are now widows and widowers; they can feel isolated, depressed, alone and anxious. Each individual's ability to cope with the challenges of illness and ageing is complicated by the suffering, loss and deprivations they experienced during the Holocaust. One son reported: 'My father had to be put on a feeding tube because he couldn't swallow. He kept asking me – Why are they doing this to me? They're starving me like the Nazis did at Auschwitz.'

Seventy per cent of care in the home in the UK is now carried out by private companies. All councils have a duty of care to appoint only those private companies whose record of competence they can verify. In the UK the choice of those who will care for the elderly has recently been severely criticised: the NHS has had a special kind of auction to appoint organisations which will care for the elderly, particularly those with dementia, and those involved were asked to reduce their costs and then participate in reverse auctions, where bids are driven down, not up – the lowest bid being the winner of the contract. To choose a carer on such a criterion seems immoral, and one company that won such an auction had the contract removed a few weeks later as their caring was so poor. To ensure that proper care is carried out, spot checks on their performance should be mandatory. Councils must also insist that carers are given professional training. Many carers are young, inexperienced, paid low wages and left to cope virtually on their own. Whistle-blowing on bad employers must be encouraged.

Concern is also growing over the expansion of home-based care for the elderly after examples of neglect have been reported. Poor standards and numerous breaches of regulations were shown in an investigation by the BBC's *Panorama* programme. There was evidence that some old people were being left alone for many hours; in one case, an 89-year-old woman was left for 24 hours before being found by her son lying in excrement. It seems that paid-for carers are not allowed enough time to give adequate care to vulnerable older people, and support services do no more than provide the minimum standard of care that they can get away with.

One suggestion is that neighbours should be paid to look after the elderly, as they already know and are friendly with the person and could provide personal care and also meals. Britain would do well to follow the French example and pay people who take time off to look after a dying relative or partner. Charities describe the care system as being in 'crisis' and one that will be dealt with head-on only when society accepts that the needs of the elderly are as important as those of newborn or disabled children. The number of over-85s will reach 4 million in the UK by 2051; for many of these a care home or nursing home will probably be the ultimate destination.

Many dislike being in a care home and so avoid communal areas and spend most of their time in their rooms. Talking about dying is taboo. There is some concern that families do not visit inmates as often as they might. There is a tradition that pupils from schools visit lonely old pensioners both to talk and to help with housework. This may not continue as both pupils and pensioners may have to be officially vetted before such regular visits in order to avoid any abuse. This is a great shame, as even when a fish tank is brought into a nursing home the patients crowd around it and conversation increases. There is a test programme in Toyota Memorial Hospital in Japan where robots monitor and interact with patients, and elsewhere 'companion robots' are being developed which, it is claimed, could lead to some improvement in Alzheimer's patients. Norway is also considering the use of robots in the future to combat staff shortages.

Caring for the severely ill elderly can be difficult and stressful. After two strokes, one patient was reported to be unable to even hold a child's plastic drinking cup, and is also unable to reach the buzzer to summon staff. While his room at the home is clean, tidy and bright, there is nothing to occupy his mind as he cannot see or hear well enough to read or watch the TV, and conversation is difficult. His only stimulation comes from family visits, but they can be days apart and upsettingly brief. He shouts and cries as his frustration surfaces, which only adds to the family's feelings of guilt and helplessness and increases their desire to cut the visit short.

Sir Michael Parkinson, who was appointed by the government in 2008 to promote dignity in care homes, described some homes he had seen as 'little more than waiting rooms for death' and said he had been appalled by letters members of the public had sent him. One had written saying her mother had been left naked, covered in urine and in full view in a side room at a hospital, while others had complained of patients' use of alarm bells being ignored for so long that they soiled themselves. Older people in hospitals and care homes are being left without enough food and drink in incidents that are 'absolutely barmy and cruel beyond belief'. Parkinson said staff and managers blamed bureaucracy for stopping them delivering more dignified care.

About 20 per cent of deaths in the UK occur in nursing or care homes. Often understaffed and with underpaid and poorly trained employees, many nursing homes endanger the lives of their patients. Those living in nursing homes receive poorer care than those living at home, because they are not given beneficial drugs; there is poor monitoring of chronic disease and overuse of inappropriate or unnecessary drugs. No study has examined the overall quality of care given to elderly patients in UK primary care, or has judged the quality of care against agreed, explicit standards in patients living in nursing homes compared with patients living at home. An example of problems is the case of a 94-year-old who was given notice to quit after she complained about the executive's dog, which kept leaping up at her. She is frail, has restricted vision and arthritic limbs, and she is unable to dress or clean herself.

Patients with dementia and related mental health problems need an enriched psychosocial environment which meets their need for human contact and personal growth. Over 820,000 people in the UK live with Alzheimer's and other dementias, costing the UK economy £23 billion per year – more than cancer and heart disease combined. Dementia research is severely underfunded, receiving 12 times less support than cancer research, even though around one third of us will suffer from some form of dementia before we die. There is as yet no basic training for healthcare professionals on how to understand and work with people with dementia. Dementia affects not just its immediate victims but also other residents forced to share their final years with sufferers. And it places a huge burden on the staff who have to look after the patients' most intimate needs. The traditional cosy view of an old people's home has long gone.

More than 2 million older people over the age of 65 in England have symptoms of depression, but according to a report by Age Concern the vast majority are denied any help. Falls by older people in nursing-care facilities and hospitals are common events that may cause loss of independence, injuries and sometimes death as a result of injury. Effective interventions are important and will have significant health benefits. The prescription of vitamin D reduces falls, as may a review of medication by a pharmacist.

Yet thousands of doctors specialising in the care of older

people believe the NHS is institutionally ageist, a survey suggests. One third of doctors thought that the old should not get various kinds of surgery as it would not last them very long. Another survey found that almost one in three nurses would not trust the NHS to care for an elderly relative who was malnourished. Old people are left hungry on wards. Discrimination against the old also involves, for example, long waits for hip and knee replacements which people over 65 need most. A British Geriatric Society survey of 200 doctors found that more than half would be worried about how the NHS would treat them in old age. Dr Blanchard, who works in psycho-geriatrics, told me that there is little evidence for ageism in his area, but what will probably happen with the coming financial cuts is that his patients will be put back into general psychiatry, and this may lead to discrimination as the cost of their care is compared to that of younger patients.

Symptoms such as frailty, wandering, agitation, falls, and lack of motivation and appetite, create a huge burden for both patient and carers. A businessman, Gerry Robinson, used two BBC TV programmes to explore care homes for old patients with dementia. The results were distressing. Inmates can spend hours without any contact with anyone. A woman cried out for help for half an hour before anyone came; in another case an alarm cord was out of reach. Staff morale was very low. Robinson argued that engagement was essential for the inmates and the system needed to be reorganised. Person-centred care is essential, care which is tailored to meet the needs of the individual rather than those of the group or the staff.

Dementia sufferers are being 'drugged and robbed' by a system where they have to pay five times their pension for poor-quality care, according to the Alzheimer's Society. More than 100,000 sufferers are being given the wrong drugs, which actually makes their condition worse, at a cost of over £60 million a year. Only about one fifth of nurses working with people with dementia receive any or enough dementia training, and almost all nurses said they found working with people with dementia very or quite challenging. A typical negative story describes staff leaving a sign next to the bed of a patient with dementia telling her: 'You are not well, you need to stay in hospital. Just sit there, rest, relax and don't bang the table.' She did not understand and could not remember anything for longer than a few seconds. Dementia patients occupy a quarter of all hospital beds and are staying far longer in hospital than may be necessary. Not only does this cost the NHS hundreds of millions of pounds, but the majority of people with dementia leave hospital worse than when they arrive, and a third enter a care home, unable to return home.

The family of an Alzheimer's sufferer have won a legal battle to reclaim more than £100,000 in care-home fees that the local NHS trust had refused to pay because it claimed that her condition was not health-related. Health authorities had ruled that the patient, who died aged 74, did not qualify for NHS funding because her condition was deemed to be a social, rather than a health, problem. As a result, she was forced to sell the home that she had lived in for 30 years for £170,000 to pay for her £600-a-week nursing home fees.

The practice of over-prescribing medication is based on the assumption that it is the natural process of ageing for the quality of health to decrease, and therefore there is no point in attempting to prevent the inevitable decline of old age. Such differential medical treatment of elderly people can have significant effects on their health outcomes. However, free influenza immunisation is being offered to everyone aged 65 and over, and routine breast cancer screening is being extended to women up to and including the age of 70.

A study in Newcastle of 85-year-olds gave a positive set of results for medical care by the NHS. Almost one third of the sample had attended outpatient clinics in the three months before the study. In the previous year, 20 per cent had had at least one overnight stay in hospital, spending, on average, seven days in total over the stays. Almost all of the sample had seen their general practitioner within the past year. Perhaps the most striking findings were the low levels of disability of people living in institutional care, and positive self-rated health despite high levels of disease and impairment. Although women were more likely to survive to age 85, they were more likely to be living in institutional care, to have a higher total disease count and higher prevalence of many diseases.

A three-month undercover investigation at Brighton's Royal Sussex County Hospital by the BBC's *Panorama* showed how hospital care can fail the elderly. In one scene a patient is left to die on her own; another patient is left waiting hours to go to the toilet; another was left screaming with pain as she had not been give her medication for hours. Margaret Haywood, a nurse with more than 20 years' experience, agreed to go undercover for the *Panorama* programme, wearing a hidden camera while working as a nurse to fill a short-staffed ward at the hospital for 28 shifts on an acute medical ward. She found that none of the patients had a care plan. She was then struck off from working as a nurse, but this was later reduced to a one-year caution.

An environmental factor, cold, has been responsible for the deaths of many old people in their homes. The average energy bill has increased by 80 per cent since January 2003, and an energy bill of £1,027 would absorb 16 per cent of the income of a single pensioner. In the winter of 2004/5, more than 30,000 people over 65 died from cold-related illnesses in England and Wales, and there were some 16,000 excess winter deaths among the over-75s. Because nearly half of pensioners will cut heating in winter for financial reasons, some 5 million over 60 will get cold shock. The UK has a higher number of winter deaths than in colder European countries, despite the government's winter fuel payments.

It is very unusual for the cold to kill people directly, and in the main these deaths are from respiratory or cardiovascular ailments. Deaths may also result from heart attacks, strokes, and bronchial and other conditions, and may often occur several days after exposure to the cold. The elderly are more vulnerable because of various illnesses and of course, the failure to warm their homes. Age Concern estimated that 250,000 older households have been pushed into fuel poverty by price hikes. Many pensioners have to choose between eating or heating.

Paul Cann, from the new united charity Age UK, says:

> To deliver consistent and decent quality healthcare for older people, dignity must be at the heart of the NHS Reform Bill. On every occasion and in every health setting, older people should be alleviated from discomfort and pain, given help to make choices and treated as individuals not numbers.

14
Adapting

'The tragedy of old age is not that one is old, but that one is young'
– Oscar Wilde

Men and women in the developed world typically live longer now than they did throughout history, an increase from about 25 years 2,000 years ago to around 80 at the beginning of the twenty-first century. The increase has been mainly due to the advances in medicine and biology which have given us vaccines and antibiotics, and the development of sanitation systems, as well as better lifestyles and better nutrition. All of these have been successful in preventing infectious and parasitic diseases causing premature deaths. But we now need to understand the implications of the increase in age of the population.

At present just 11 per cent of the world's population are over 60, but in developed countries they will be one third of the population by 2050. In rich countries one in three individuals will be pensioners and one in 10 over 80. In some countries in the West the over 65s are the fastest growing age group. Those over 80 in the world are expected to increase to 4 per cent by 2050, four times more than now. Current estimates are that 700,000 of those in the UK at present around 25 years old will live to be a hundred. Moreover, half of babies being born now will reach a hundred thanks to higher living standards. But our bodies are still wearing out. Children will be outnumbered by those over 75 in what some call the Zimmer-frame society. How will society adapt?

A majority view is that an increase in numbers of older people would make no difference to safety, security, standards of living, health or access to jobs and education. But one third of the public think life would be worsened by an increase in the older population because it would have negative economic effects. It is worth noting that by 2050 it is estimated that more than one third of voters in the UK will be over 65, and since the old vote more than the young, they could wield much power and vote for much costly support in their old age.

David Willetts, the current science and education minister, has argued that the baby boom of 1945–65 produced the biggest, richest generation that Britain has ever known. Today, at the peak of their power and wealth, baby boomers, he claims, run our country; by virtue of their sheer demographic power, they have fashioned the world around them in a way that meets all of their housing, healthcare and financial needs at the expense of their children. Social, cultural and economic provision has been made for this reigning section of society, whilst the needs of the next generation have taken a back seat. But it is the old who will have the greatest impact.

Some analyses suggest that improvements in health and longevity have resulted in enormous gains in economic welfare. One estimate of the economic impact of post-1970 gains in life expectancy suggested that they might have added as much as 50 per cent to the GDP of the US. Even so, there are profound economic problems to be tackled.

The *Economist* has called the economic effect of an ageing population a slow-burning fuse. It claims that age-related spending by a country like the UK will in the future be more serious than the recent recession. The elderly require money for pensions, health and care. The age-related spending by the government in the UK is already about £7 billion, less than 1 per cent of GDP, but it clearly needs to rise. Already there are billions of pounds of benefits for the old which are not claimed. The last years of life can cost tens of thousands of pounds and this increases with advanced age. In many cases the cost of the last year of life is more than all that has been spent in earlier years. Forcing older workers to retire cost the UK in 2009 an estimated £3.5 billion in lost economic output. Pension plans, social security schemes and notions of the length of working lives will need to undergo major reformulation.

The small increases over time in average life expectancy that lead to very large increases in the size of a population are prompting many arguments. Reproductive practices might have to change in order to keep the population from becoming too large. In regard to the escalating costs of looking after the old, one group in the US has suggested there should be cuts to protect the young. It has been argued that any technological advances in life extension must be equitably distributed and not restricted to a privileged few. There are suggestions that the UK will become a giant residential home, with the young looking after the old. With family ties weakened by increased mobility and rising divorce, in the future the elderly will be less likely to be married or co-habiting, and more will live alone; and since they do not want to leave their homes it will hard for the young to find one to buy.

Such concerns were presented by Jeremy Lawrence, writing in the *Independent*, who described the ageing population as the greatest threat to human society: 'No invading army, volcanic eruption or yet undreamt of plague can rival ageing in the breadth or depth of its impact on society . . . The impact of this transformation will be felt in every area of life, including economic growth, labour markets, taxation, the transfer of property, health, family composition, housing and migration. And the "demographic agequake" is already under way.'

The countries with the oldest populations – that is with the highest percentage over 65 – are Monaco, then Italy and then Japan. The median age of the world population will, in the next 40 years, go from 28 to 38. The United States is on the brink of a longevity revolution. The elderly comprise 12 per cent of the US population, and their number is projected to almost double between 2005 and 2030, from 37 million to 70 million. By 2030, the proportion of the US population aged 65 and older will double to about 71 million older adults, or one in every five Americans. The far-reaching implications of the increasing number of older Americans and their growing diversity will include unprecedented demands on public health, ageing services and the nation's healthcare. Although there may be unjustifiably pessimistic views of what is before us, an ageing population does present severe problems. But there are also great advantages.

An analysis of the consequences of increasing life expectancy must include the economic implications of changed population age structures, especially changes in the ratio of those in the labour force to those outside it, mainly children and elderly, which is known as the support ratio. In the early stages in which life expectancies rise, the proportion of the population within the labour force age range rises significantly, providing a substantial boost to economic growth. However, as people live longer those extra years will have to be financed, and there are only a limited number of ways to do this. These include working longer, increasing social security or other taxes, increasing immigration, and reducing consumption.

In order to reap the economic benefits of longer lives, some of the extra years of life will probably have to be spent working productively. In the UK one in five over the age of 55 can expect to work till they are 70 and even older. This is because of their limited finance. At 55 many still have a mortgage and less than £2,000 in savings. There is considerable variation in the EU about attitudes to the extension of working life 9 out of 10 in the Netherlands, Denmark, and Finland are positive, but in Greece, Portugal, Spain and Hungary they are not at all keen. It may be sensible to compensate younger people for working when they are older by allowing them to work fewer hours per week over the whole course of their lives.

The number of people supporting pensioners is decreasing, and it will go from four workers for each pensioner at present to two workers for every pensioner in 2050. A major problem of the ageing of societies is that people are having fewer children. The current global average is 2.6 children per woman, while in rich countries it is 1.6. One reason is that women are having babies later in life. This means that in some rich countries the population is beginning to decrease. In ageing countries, the economy can shrink as more retire and there are fewer young to take their place, and the older workers may be less productive. Japan has a very low ratio of workers to pensioners, just three to one, and by 2050 the number will probably halve. Such changes can reduce economic growth significantly. In some countries immigration is filling this labour gap; another solution is to encourage people to have more children, as has been done in France and Japan. There is also the problem of finding enough young adults for the armed forces.

Pensions raise serious economic issues. The official retirement age for most developed countries has remained the same even though the population is ageing. Many have even retired before the official retirement age. Retirement pensions are the largest component of age-related spending, and the cost of state pensions in rich countries will probably double by 2050 and reach more than 15 per cent of GDP. Payments to the pensions of retired civil servants in the UK which are based on final salary are very expensive – the pensions can be as large as two thirds of final salary. Each one-year increase in longevity increases costs £1.3 billion a year.

Part of the problem is that men can now look forward to between 14 and 24 years in retirement, much more than anticipated. Older workers also want a less onerous workload. The UK government's decision to abolish the compulsory retirement age, currently 65, wll increase the number of the workforce. Retiring later will help financially, but will not the young then be deprived of senior appointments? It is estimated that unless urgent action is taken there will be a £6 billion hole in the funding of social care within 20 years. Free care at home could cost the state more than £1 billion. A very big financial hole is being opened up.

The Big Question: 'Is the world's population the wrong age?' In some countries there are too many old people, while in others the population is very young. While the developed world is facing up to the challenges of an increasing number of elderly citizens, some developing countries are facing the strain of populations in which a third of the people are under the age of 15. Neither situation is ideal, but the challenge is to find ways of adapting to cope with the economic demands of differing population ages.

According to a UK survey, most individuals approaching old age had not yet thought about how they would be cared for. I myself am guilty of this. At least 13 million, according to the government, are not saving enough to retire on a decent income. Some 70,000 people have to sell their homes each year to cover the cost of long term care. When asked about the supposed worries they would have when 75, about half those aged 45 to 75 were rightly worried about money for long-term care. At 65, the government says, a woman can expect to face average care costs of £40,400 and a man, who will not live as long, £22,300. Those with severe health problems such as disability and Alzheimer's disease and and other forms of dementia are by law entitled to free NHS care, but ageing illnesses makes self-support increasingly difficult. Plans for nearly half a million needy elderly to remain in their own homes would cost £670 million a year, and in current economic circumstances such figures are vulnerable.

The health of the elderly remains a major issue. Health

spending on the elderly in the EU is about one third of the total health budget. In the UK more than 40 per cent of the NHS budget is currently spent on people aged over 65. The estimated number of people in the UK living with the effects of stroke, which mostly strikes people over the age of 60, will rise significantly, and the number of people with dementia will increase to around a million in 2025 imposing a major burden on social services and families. The cost of looking after them has been predicted to increase to £35 billion. The number living with coronary heart disease, osteoporosis, osteoarthritis and age-related macular degeneration of their sight will all increase dramatically. Treatment of all these illnesses is, to put it mildly, expensive. Currently those over 65 consume one third of all drugs but are only 14 per cent of the population. The NHS will face increasing demand for its services to the old. To go further and increase the quality of services will require an additional 3 or 4 per cent increase each year.

According to the Institute of Medicine report in 2008, the elderly in the US account for more than one third of all hospital stays and of prescriptions, and more than a fourth of all office visits to physicians. The average 75-year-old American has three or more illnesses and takes at least four medications. Delivering optimal geriatric care has become a costly medical and ethical priority. Medical and nursing schools are, it is claimed, training far too few doctors and nurses on how to care for the elderly. At the same time, other workers, such as nurses' aides and home health workers, remain under-trained and underpaid, the experts say. The number of doctors specialising in geriatrics has been falling. The US government has recently announced a new initiative to assist with housing costs for elderly and disabled people.

Despite the best efforts of any government, the cost of the elderly is going to leave a gap in funding. Spending on care for the elderly will have to double over the next 20 years to cope with a surge in the numbers of sick and disabled old people. It was previously believed that the amount of time that pensioners spent being sick or disabled would remain constant or even shrink with the help of medical advances, but this is unlikely. Instead, many of the extra years will be spent being unwell and in need of care. Increases in the number of years of good health have not kept pace with improvements in total life expectancy. The number of sick elderly people, or those with disabilities, will increase by around two thirds over the next 20 years. The costs are frightening.

Around 700,000 people in the UK spend more than 50 hours a week caring for a relative, according to Carers UK. A lack of facilities means one million family members already take time off work to care for aged or disabled relatives, while another six million take some ad-hoc responsibility for caring. By 2033 the number of people aged 85 and over is projected to more than double again to reach 3.2 million, and to account for 5 per cent of the total population, and so the number of carers needed will have to soar as well. Britain faces a care time-bomb within seven years, with the number of elderly needing full-time help outstripping the number of carers. Free personal care at home is available in Scotland but not in the rest of the UK.

Public transport systems, especially fixed-route bus services, face important challenges in meeting the needs of the elderly for convenient transport. There is a need for wider pavements to make the roads and streets safer for older pedestrians. This could include dedicated pathways for electric wheelchairs, improved access points to public transit and commercial areas, along with special ramps or expanded parking spots for the ageing population. Yet more costs.

In China, meanwhile, an enormous population is also ageing rapidly. By 2050 about one quarter of all Chinese will be aged over 65. This is one of the consequences of the country's 'one couple, one child' family planning policy made 30 years ago. This puts China in a difficult situation. With a population of 1.3 billion – the world's largest – and one that is set to peak at 1.5 billion in 2026, the authorities cannot afford to relax their tough birth control policies. But without more younger people, who is going to support the hundreds of millions of elderly? The percentage of elderly is projected to triple from 8 per cent to 24 per cent between 2006 and 2050 – to about 320 million old people. It is almost certain now that China's generation of only children will find themselves as adults trying to support two retired parents and four ageing, and possibly ailing, grandparents. Officials are already talking anxiously about the 4-2-1 phenomenon. How are those costs to be borne?

In Japan in 1950 there were 97 centenarians and in 2008 there were 36,726. Reaching a hundred is no longer the miracle it used to be. It has one of the oldest populations; there are just 3.4 people working for each one over 65, and by 2050 the number will be just 1.3, worryingly few. Can the pension books be managed as fewer children are being born? Japan has the world's fastest-ageing population and is very short of care for the elderly. It was estimated that 100,000 new carers will be needed by 2010; as the recession has caused much unemployment and 3 million are jobless, an attempt is being made to train many of these as carers for the old.

What of the future? The charity Age UK has an agenda for later life. Age UK is in partnership with over 300 local Age UK organisations and the overall budget is around £400 million, of which about £250 million involves these local organisations. They get about £50 million from insurance, about another £50 million from their shops selling clothes and mobility devices like stair lifts, and around a final £50 million from donations. I spoke to their chief executive, Tom Wright:

> Age UK has a very simple vision, which is a world in which older people flourish. Our purpose is to help to improve the lives of people in later life. We believe that an ageing society is one of the biggest challenges globally, but there is also an enormous opportunity as an ageing society brings with it decades of wealth, experience and knowledge which at the moment is not always fully appreciated. That is the challenge for us, being the largest organisation focusing on later life in the UK, and we have set out to deal with it. There are five important areas: money matters, wellbeing, care at home, work and education, and finally leisure.
>
> At present the most important is care. Living in your home is an area the government have not fully got to grips with. We campaign on all these issues and it is based on sound research and we have several hundred forums with older people around the country where we continually get feedback on issues that are affecting older people, day in, day out. And then we employ some of the experts in the different disciplines which allows us to go to

the government and business with the best knowledge about the issues and to work with them to develop solutions.

Many of the issues are interrelated but care and wellbeing is a very significant one. Wellbeing affects health and evidence from the research we do shows that activity helps in later life and prevents cognitive decline and depression, and can lead to less care having to be given. The new government has picked up on many of the manifestos and ambitions that we have sent to them. Before the election we laid out for all the parties what the key priorities are, and many of these have been retained by the new government. All the indications are that it includes getting rid of compulsory retirement at the age of 65. There will also be prevention of erosion of state pensions. They are putting forward a care commission to look at these issues.

We do not focus on the illnesses like dementia and cancer that affect later life, but on the quality of life. We do work on cognitive decline and how mental abilities change. There is a study by the University of Edinburgh on a cohort of 1,000 in Scotland starting in 1947 from the age of 11. We do a lot of work on incontinence, which is an area that others are less likely to focus on, particularly infections of the bladder which are key contributors to incontinence. The tests for incontinence are not sufficiently given, and incontinence is also linked to dementia. We are concentrating on improving later life rather than extending life. We do not take a strong position on euthanasia one way or another.

There is a sense that the old are put in a box labelled 'old' or 'aged' and there are many cases in medicine where the old do not get the necessary care because the condition is claimed to be age-related. A person going to a doctor, for example, and saying that he has a bad knee, the doctor would say that's because you are old, and the person would say that his other knee is fine. With regard to care in the community, younger people get much higher budgets than older people. There is at times a lack of respect for the experience and knowledge of the old.

Another important issue is that there are over 2 million older people living in poverty. And many would be pulled out of poverty if they took the benefits they were entitled to. This is partly due to the complexity of the process and the forms that need filling in. Half of those over 75 live alone and do not have someone to help them with this. We would like auto-enrolment which means that the government would automatically give you your entitlement. It is a complex area.

14
Ending

'Old men should have more care to end life well than to live long'
– Captain J. Brown

In 1965 The Who sang 'I hope I die before I get old.' When young, one thinks little about dying, but when old it is almost impossible to avoid it.

It is not necessarily frightening, but something for which one must be prepared. We elderly are constantly asking ourselves what makes life worth living. Are we scared of dying? A good death requires us to retain control, know when and where it will happen, have pain relief and access to good medical care. Centenarians are apparently allowed to die quickly, but the 85-year-olds are not. The bioethical critics of anti-ageing research and radical life extension lament the fact that 'we' are unable to accept death. Bioethicist Daniel Callahan argues that we must learn to accept the idea of a 'natural lifespan', one that might reach its conclusion sometime around the age of 80, for then surely we have more or less had adequate time to enjoy our creative capacities, raise children and experience what life has to offer.

How much is the chance of dying dependent on your age? Unsurprisingly, about 80 per cent of all deaths are of people aged 65 and over. Two thirds of deaths in England occur in people over 75. Taking all diseases together but not including accidents, in the developed world like the UK the death rate at 80 is 500 times greater than at 20. What causes this difference is not simply the ageing of our cells but many time-dependent processes. For example, with cancer there are many stages to be gone through, which take time. The same is true for many other illnesses, such as those affecting the blood system and the heart. It takes time for the vessels to become blocked. Fewer than one in 20 want to die in hospital, but nonetheless one in five do. In the UK, of those who were aged 65 and over when they died, about three quarters died in a hospital or in a care home. It is generally accepted that a supported death is preferable at home. Thinking about death may not be comfortable, but supporting people in the closing months and weeks of their lives should lie at the heart of the health service's mission. The final year of life also accounts for a very high proportion of the costs of many people's lifetime healthcare.

Several social factors can influence when one dies. Those who expect to die soon do in fact do so, compared to those who have longer expectations. Individuals who are between 50 and 59 years old and from the poorest fifth of the population are over 10 times more likely to die sooner than their peers from the richest fifth. This and other key findings emerge from the latest results of the English Longitudinal Study of Ageing (ELSA). And in the USA, elderly Americans with low education levels are more likely to die from serious illness, suffer disabilities and experience a lesser quality of life than their better-educated senior citizens. They also recover more slowly from hospitalisation. The reasons for poor health among these people may have to do with higher levels of hostility and hopelessness, and being ill equipped to maintain health.

How should one prepare for death? Should people, as Dylan Thomas asked, 'Rage against the dying of the light' or go gentle into that good night? Death anxiety is a common predictor of negative attitudes to ageing. For some it is the anxiety about the process of dying, and for others the uncertainty of what and where it leads to. There are many end-of-life decisions to be made: wills and, given the choice, where to die. It is very important that the old prepare their wills. A will is valid provided the testator understands it, and a delusionary state can invalidate a will. One can also make a living will and give medical decisions related to death to someone else, such as not being revived when very ill.

Open discussion may not be possible in the final stages. We need to avoid having to make last-minute decisions if possible – death requires a lot of preparation. Two thousand years ago Seneca wrote, 'He will live badly who does not know how to die well.' This is so much more relevant now that we are living so much longer, and death can come much more slowly. At the age of 80, Churchill said that he did not mind dying as he had seen everything there was to see. Virginia Ironside also has a positive view:

> Death, like grandchildren, is one of the extraordinary new and exciting perks of old age. Over 60, it's time to get acquainted with it. No use dreading it or being frightened by it. People are always wringing their hands when their friends die but frankly, what did they expect? That they'd live for ever? What you don't want is to let death take you by surprise, or you're going to be like people who find that when the car comes to take them to the airport for their holidays, they have forgotten even to start

packing. Visit the dying. Look at dead bodies. Write your will. Face up to it. It's an adventure.

Carl Jung wrote that 'it is hygienic . . . to discover in death a goal to which one can strive; and that shrinking away from it is something unhealthy and abnormal which robs the second half of life of its purpose.' It is suggested that being reminded of our mortality can be a stimulus to a spiritual awakening. Barry Cryer from BBC Radio's *Sorry I Haven't a Clue* was asked when 74 if he feared death and replied, 'No. I'm getting old, but it's unreal. It's meaningless. I don't mean that I'm kidding myself, but age is just a number. I'm not afraid of death.' Woody Allen said, 'I just don't want to be there when it happens.'

Does ageing itself lead to death? A peculiar and quite difficult problem is whether anyone actually dies from old age, and the answer seems to be no. There is almost always a good medical explanation for anyone who dies when very old in terms of the abnormal behaviour of their cells and organs that gives rise to a well-recognised illness. However, death certificates can give the cause of death as 'old age' in the UK, and some do. This fits with the countless grandparents that are claimed by their relatives to have 'died of old age'. They attribute an elderly person's death to old age because no other obvious explanation emerges. But in the USA, nobody has died of old age since 1951, the year the government eliminated that wording on death certificates. There is a limit to the human lifespan, but in many cases elderly deaths are pinned on old age simply because no one looked very hard for the true cause.

Autopsies of 40 centenarians who died at home, and seemed to be quite healthy, found a cause in every case.

The very old are often felled by an infection or ruptures in the aorta, the major vessel that moves blood from the heart, and more than one factor often triggers death. While no one dies of old age, ageing does lead to the inability to deal with a disease that may be partly due to the ageing process itself.

Some two thousand years ago Marcus Aurelius commented: 'Mark how fleeting and paltry is the estate of man – yesterday an embryo, tomorrow a mummy or ashes.' When does death begin? Perhaps at birth.

Even if an illness is ultimately to blame for a death, it is often preceded by a downward spiral that renders a person particularly vulnerable to dying. More than two thirds of those who die over 85 are women who are suffering from lingering illnesses. Men may avoid going to the doctor and are accident prone. Falls are a major cause of disability and a leading cause of death from injury in people aged over 75 in the UK. The main causes of death over 65 – heart disease and cancer – are the same for both sexes.

The causes of death have changed. The leading causes of death in the US in 1900 were infectious diseases, but by 1940 they were heart disease, cancer and strokes, and by 2004 heart disease and cancer. Figures were quite similar for those in the younger age groups. Since the 1980s there has been a decline in death rates for the over-65s but none for younger groups such as those between 50 and 64. As we have seen, 80 per cent of deaths are of those over 65 years. More people now die as a result of chronic illnesses such as heart disease, vascular disease including stroke, respiratory disease and cancer. Older people may suffer from several conditions at the same time, which may make it difficult to determine the main cause of death. There seems to be little genetic influence on the age of death, as identical twins still differed by 14 years, while non-identical differed by 19 years. But, as discussed earlier, the APOe4 gene, which is linked to getting Alzheimer's, is often absent in the very old.

In some cases older people may be transferred into a hospice for terminal care, which helps people to live as actively as possible after diagnosis to the end of their lives, however long that may be. In spite of the fact that about 20 per cent of older people die in care homes there has been little emphasis on the needs of older people dying in a care-home setting or how well these are met. In order to provide good care for people at the end of their lives, care-home staff need external medical support, particularly from GPs. Without this support, symptom relief may be poor and a resident may have to be transferred to hospital or hospice to die. Although this may be appropriate in some situations, there are inappropriate transfers from care homes. This can be traumatic for the older people and their families. The factors which can influence this process include a lack of forward planning, no knowledge of the older person's preference, poor relationships with GPs and a shortage of resources in the care home. Several commentators have suggested that it is only by removing the taboo of the discussion of death, throughout all stages of life, that a better understanding of the realities of dying and death, better communication skills and ultimately better service provision will be delivered.

Caring for someone who is dying can be distressing and demanding for carers, including family, and in the case of people with chronic illnesses this may extend over a long period of time. If the demands on carers become too great the arrangements for the care of the dying person may break down. In practical terms this means that the dying person may not be able to die at home, even if that is their wish. Carers need information and both practical and psychological support. Professionals need to coordinate the support they provide to both the dying person and to carers. Although the gap in life expectancy between women and men has narrowed, women are still more likely to outlive men. Bereavement and coping alone are thus much more common experiences for women than for men and are likely to remain so. Not untypically, one woman whose husband was in a home experienced guilt, sadness, shame, love, and resentment that he was still alive.

While three quarters would prefer to die at home, people who are recently bereaved, and therefore have experience of death, are slightly more likely to prefer in-patient hospice care. There is a clear inverse relationship between where people say they want to spend the last period of their lives and where they actually die. There is significant variation across European countries as far as place of death is concerned, with some of the highest rates of deaths in hospital occurring in England and Wales. This suggests that the organisation of services plays an important role in determining the options that people can consider.

Health professionals must ensure that older people who want information about their diagnoses and prognoses should be given this in a sensitive and appropriate manner. Older people who wish to do so should be given the opportunity to discuss their dying and death. People at the end of their lives should be given a choice of where they wish to die, and how they wish to be treated, regardless of diagnosis. If older people are not in a position to make such choices, because of mental incapacity, decisions should be made in their best interests and in consultation with people who are close to them.

The old do commit suicide, but are not more involved in suicidal thinking than the young – thinking about suicide is most common in the 25–44 age group, though feelings of hopelessness are common in the elderly. Ageist attitudes can also be found in the language used to describe the end of older people's lives, and in the description of the suicide of older people, which tends to be portrayed as heroic rather than as tragic. Indirect forms of self-destruction involve refusing treatment and food and are most common in care or nursing homes. Self-harm usually involves poisoning. Older men commit suicide more than women. Research has found that over half of older people who take their own lives were experiencing depression at the time of death. Men over the age of 75 have the highest suicide rate amongst all groups. For some, death can be welcome if the suffering is severe, and there is no hope of improvement. A recent study has identified illnesses which may increase the suicide risk. Depression, bipolar disorder (manic-depressive illness) and severe pain – but not dementia – were associated with the largest increases in suicide risk. However, several other chronic illnesses including congestive heart failure, and chronic lung disease, were also associated with an increased risk for suicide. The researchers also found that treatment for multiple illnesses was strongly related to an increased risk of suicide, and that most of the patients who committed suicide visited a physician in the month before death, about half of them during the preceding week.

In his book *How We Die* Sherwin Nuland gives a detailed description of a friend of his with Alzheimer's and the serious effects it had on his wife. When he passed away she wrote: 'And when he died, I was glad. I know it sounds terrible to say that, but I was happy when he was relieved of his degrading sickness. I knew he never suffered, and I knew he had no idea what was happening to him, and I was grateful for that.' A friend told me about her 96-year-old mother and her 92-year-old husband. He was effectively blind and had difficulty going upstairs. Her eyesight was poor but she devoted herself to looking after him. She said he had so changed that he was now quite unlike the person she loved, and his peaceful death would be far the best for both of them. This, thankfully, has now occurred. Death can be a relief. This leads us to consider assisted death, or euthanasia.

Instead of suicide, would not voluntary euthanasia be much preferred? Euthanasia is the intentional ending of life by a painless method for a person's alleged benefit. It is usually assumed that it has the individual's agreement, even wish. There are some subtle differences which can have legal implications. Voluntary euthanasia is when death takes place with the patient's consent, and is different from occasions when the patient has neither requested nor agreed. There is also a distinction between active and passive, the former involving lethal injection, and the latter simple medication or the withdrawal of medication. Assisted suicide is when the patient takes the last step and another person provides the means of bringing about the end of their life. Under current UK law euthanasia is classed as murder, but recently cases of voluntary euthanasia have not been prosecuted. Assisted suicide is legal in

Holland, Switzerland and the states of Oregon and Washington in the USA. It is hard to for me to accept the ban on voluntary euthanasia or assisted suicide for the terminally-ill elderly – I cannot accept the reasons that are given.

In ancient Greece and Rome, euthanasia was an everyday reality for many people who preferred voluntary death to endless agony. This widespread acceptance was challenged by the minority of physicians who were part of the Hippocratic School. The ascent of Christianity reinforced the Hippocratic position on euthanasia and culminated in the consistent opposition to euthanasia among physicians. Proposals for euthanasia revived in the nineteenth century with the revolution in the use of anaesthesia. In 1870 Samuel Williams first proposed using anaesthetics and morphine to intentionally end a patient's life. This led to much discussion within the medical profession, particularly as to how much autonomy should be given to doctors.

The elderly may be exposed to backdoor euthanasia under the Liverpool Care Pathway. With patients deemed to be terminally ill, and if they think the patient is near death, doctors can withdraw fluids and drugs, so the patient, while on continuous sedation is allowed to die peacefully. This seems an attractive procedure but there is some concern about this process, as when under sedation improvement in the patient's condition cannot be detected, and the doctors involved are not geriatricians. The decision to withdraw treatment is clearly a complex one, but doing so can greatly reduce the suffering of both patient and relatives

Geronticide – involuntary euthanasia – is the modern term to describe the deliberate killing of the elderly because they are old. Julius Caesar is reported to have said that the Romans killed the old who wanted to die, as society was orientated to fighting, and to die of old age was shameful. It was common among some non-industrial societies, and the choice of some agricultural or nomadic communities with inadequate resources was to sacrifice the old. Not infrequently relatives and friends regard these acts as deeds of mercy, and the aged sometimes welcomed and demanded them. Hunter-gatherers are less likely to care for the old when they are less able to gather their own food. Australian Aborigines buried the old in a hole until only the head showed, and let them die. It was a custom among the Dinka tribe in Sudan to give live burial to the old. Bushmen in Africa valued the old for their knowledge and experience, but once they became incompetent they were neglected and could be put on an ox and sent to a remote hut to die. Among the Yaghan indigenous peoples of Tierra del Fuego, the old were cared for but put to death when their condition was considered, by general agreement, hopeless. The same occurred with the Koryak in northern Siberia. In some parts of Japan there was a custom of holding a ceremonial feast every three years, followed by deportation of the old to a sacred mountain to eventually die. Until recently, certain communities expelled old age people from their midst.

John Humphrys, a presenter of the *Today* programme on BBC radio, cannot forgive himself for not being able to help his father die. He listened to his old father's cries in the confines of a mental hospital. Would, he wonders, he have done anything wrong if he had helped him die – actually killed him? When interviewed, he compared the rich who wanted to end their life and were able to go to

Dignitas in Switzerland with a poor ill old lady who has no one to help her. He points out how terrible someone with severe Alzheimer's can be for a family. Thousands of people wrote to him when he described his anguish over the death of his father, and he subsequently wrote a book supporting euthanasia. In their book *The Welcome Visitor: Living Well, Dying Well*, John Humphrys and Sarah Jarvis argue that our attitudes to death and how we handle it need changing as we are living so much longer. We need to plan our death so that there is a minimum of pain and anxiety. We can sign a living will, so that if we have a bad stroke or other very serious illness we will not be revived. Since the book was published, a case before the Law Lords and new prosecution guidelines mean that relatives are less likely to fear prosecution in the future. I believe that euthanasia should be supported and it is unjust that relatives or carers who take a patient to Dignitas in Switzerland to end their lives should be liable to prosecution.

Baroness Mary Warnock, a supporter of euthanasia, believes we have the right to choose to die. This is particularly relevant to the old suffering from serious illnesses, especially if they feel they are a burden on their family. Many oppose this view and claim that one can have a good quality of life even with dementia. Patients with severe dementia may not be able to make rational decisions about death, so there could be a document a patient signs saying that when incontinent, very ill, and unable to even recognise relatives, death is preferred. Martin Amis is very pro-euthanasia: 'My stepfather died horribly. I think the denial of death is a great curse. It was a lost battle and we all wanted to assist him.' Many doctors do not support euthanasia, for while they are sympathetic and do not oppose it on ethical grounds, they do not want to be the killers. A majority of the UK public support assisted suicide.

All major religions teach that physical death is not the end, and for many older people and their families it may be important to help the transition from earthly life by performing religious ceremonies and rituals immediately before and after death. Some family members who are, for example, Catholics will insist on all possible treatments to prevent death and are totally against euthanasia. However, in spite of the right-to-life conviction, Catholic bishops have argued that it is necessary to weigh the benefits and burdens of life-saving treatments. Jewish thinking takes a similar view, and says there should not be attempts to prevent death when it is inevitable.

I am attracted to Trollope's suggestion in his book *The Fixed Period* (1884), in which a colony near New Zealand need to deal with an ageing population. They decide that anyone over 67 must die and thus be saved from the problems of old age. I once proposed we all should have a gene which ensured painless death when we were 80 and that as everyone knew about this limited lifespan, it could be a great advantage to everyone. I have now increased that age to 85.

Perceived age and looking well, which are widely used by clinicians as a general indication of a patient's health, are robust biomarkers of ageing that predict survival among those aged 70 and over, and correlate with important functional and medical conditions. So if you are told you are looking well, enjoy it for as long as you can. I find it difficult.

15

Enduring

'And in the end, it's not the years in your life that count. It's the life in your years' – Abraham Lincoln

In writing this book I have learned a great deal about the serious problems that face many of the old. In addition to the problems involved in how the old are to be cared for, I hadn't known how many of the old are so poor, and that so many need major help. I also did not know about the extent of discrimination, and why compulsory retirement is bad for so many, or how serious are the problems of of loneliness. I am nevertheless very impressed with how some of the very old cope with their age and enjoy their life. For all these problems charities like Age UK play a most helpful role and need to be supported.

I have also learned much about the biological basis of ageing, which is full of surprises, particularly the key role of evolution – we are only here to reproduce, and ageing is the result of wear and tear that is only corrected until reproduction is over. There is still much to be learned about that wear and tear in cells but progress has been impressive. There is, for example, the need to understand why germ cells do not age. Even so there are the remarkable systems in many animals whose activation can increase longevity, such as the one involving insulin signalling. Their role in human ageing is less clear. There is at present no real evidence for any way of making us immortal or significantly increasing our lifespan to, say, 150. And would we really want that unless the effects of ageing were also absent? Much more important is to find ways of reducing the effects of ageing, particularly illnesses like dementia. Politicians need to give much more attention to the problems of the old than they currently do, though the Department of Health has issued a Prevention Package for Older People. There is a strong case for there being a minister for the elderly. But for many the good news is that the government is to abolish the compulsory retirement age.

There are also major economic issues due to the ageing population of which I was unaware. For an overview I consulted Dr Richard Suzman, the director of the Behavioral and Social Research Program of the National Institute on Aging, National Institutes of Health, USA. We do not have a similar institute in the UK. His views, with which I agree, provide helpful conclusions to this study:

> Population ageing is a worldwide phenomenon. We are on the brink of an historic watershed and transformation. Within maybe five years or so, for the first time in history, people aged 65 and over will outnumber children under the age of five. One might expect that this will be true for the rest of history, and the same will hold for the 65-to-under-age-15 ratio. Over the next few decades, the older population is expected to grow fastest in low-income countries. Countries age in terms of population structure initially when fertility declines and then subsequently as life expectancy increases. Few expect fertility to ever rise to its previous levels, and there seems no end in sight to increases in life expectancy. Because of the high fraction of immigrants and their high fertility rate, the USA is a younger country than those in Europe, in terms of population age structure. Low-income

countries are ageing before they become wealthy, and China's one-child policy accelerated population ageing. Population ageing inevitably will present each nation with a welcome challenge – how to provide for people in old age once they have left the workforce. The alternative – that no one ever reaches the age of retirement – is less inviting.

At a societal level, the core issue is economic. The extra years of life, welcome as they are, need to be somehow financed. This can be done in a number of ways: people can work longer, consume less during their lifetime, save more for old age, consume less in old age; governments can raise taxes on those still earning to support retirees (along with children, the other dependant members of society), expand the economy through increased productivity, encourage high levels of immigration of working-age individuals, etc. The important ratio is the fraction in the labour force versus the fraction being supported out of their earnings (coupled with any private savings and pensions – which hardly exist in some nations). Balancing these needs in a way that allows for continuous economic growth and the well-being of future generations is what it is all about.

Leaving aside the issue of combining population ageing with economic growth and solvency, I think it is probably more important to maximise health expectancy than life expectancy. Old age is powerfully associated with physical and cognitive disability, especially among the 'oldest old', those over age 85, one of the fastest growing age groups in industrialised countries. What is important here is that the relationship between ageing and disability is plastic rather than fixed and immutable. Over a 20-year period, the prevalence of disability in the USA declined by 25 percent in the older population, though the increase in obesity may be eroding and even reversing that very positive trend. The goal of the National Institute on Aging, a component of the National Institutes of Health in the USA, is to improve both the health and wellbeing of older people. As the science of measuring subjective wellbeing improves, I would also add the maximisation of wellbeing to the mix. There are very large differences in life expectancy across both regions and social classes in the UK, with even greater internal differences within the USA (a country that since the early 1980s has lagged behind other industrial countries in life expectancy). Addressing these major inequalities should be a high priority.

In a different way, the same holds for health. The creeping obesity epidemic is likely to result in high levels of diabetes and functional disability, which will increase the demand for expensive long-term care. Efforts – or perhaps I should say, lack of effort – today could have long-term consequences for the health of future generations of elderly. Currently, for example, there are no clear-cut and experimentally confirmed ways to prevent dementia and Alzheimer's disease. The return on finding ways to prevent or delay the onset of these diseases would pay enormous dividends, both economic and in terms of wellbeing. A question I will pose, but not answer, is how governments should balance these issues against further investment in research addressing the problems. I hope that interventions that delay, prevent or remedy Alzheimer's disease will be found within a few years. I hope that more governments in low-resource countries begin to think more seriously about their demographic futures and begin to set in place policies needed for the future. Barring disastrous new diseases, I suspect that life expectancy may increase faster than many official predictions. I fear that growing obesity will counteract some of the positive trends we have seen towards lower levels of disability in the older population. At the level of both the molecular and the whole organism, we only partially understand the process of ageing. But it is not impossible that researchers may stumble on interventions that slow down the ageing process in humans without other negative biological consequences. This would have major consequences for people and societies.

Seven out of 10 people aged 65 and over believe politicians see older people as a low priority, and the former

UK health secretary Andy Burnham has said the NHS must be re-engineered to cope with the demands of an ageing population. More care is needed, and much of this could be moved into the community. Something has to be done to prevent the elderly selling their homes and using up their savings. In a scheme worth considering, everyone who could afford it would pay into a state-retirement insurance, and then receive complete cover for their problems as they aged. Age UK is challenging the government and all political parties to transform the ageing process by ending pensioner poverty, banning all forms of age discrimination, and ensuring older people can access better-quality care and support.

Will society be able to afford the high cost of care and medical needs of the elderly? Even if all of Age UK's priorities for improving life for the elderly are achieved, the underlying economic problems of an ageing society will remain here and in other countries. It is a major problem for the twenty-first century. There also has to be much more research into the ageing process to provide ways of dealing with age-related illnesses.

Writing this book has helped me to deal with my own ageing and my anticipation of death. I think it is a subject to which all of us should give much more attention. For the moment I remain looking well, but who knows for how long. In general I believe we should die before the ravages of old age really damage us. We should strongly support euthanasia for those who want it. I do not want to be one of the one in four in a care home and I would be happy to die peacefully at home when I am 85. But I may change my mind. Please keep remembering that research world-wide has shown that we are least happy in our mid-forties and happiest in our late-seventies, and even older.

*

Finally, here are the resolutions made by Jonathan Swift (1667–1745) for 'When I Come to be Old', written some 300 years ago:

Not to marry a young Woman.
Not to keep young Company unless they really desire it.
Not to be peevish or morose, or suspicious.
Not to scorn present Ways, or Wits, or Fashions, or Men, or War, etc.
Not to be fond of Children, or let them come near me hardly.
Not to tell the same Story over and over to the same People.
Not to be covetous.
Not to neglect decency, or cleanliness, for fear of falling into Nastiness.
Not to be over severe with young People, but give Allowances for their youthful follies, and Weaknesses.
Not to be influenced by, or give ear to knavish tattling Servants, or others.
Not to be too free of advice nor trouble any but those who desire it.
To desire some good Friends to inform me which of these Resolutions I break, or neglect, and wherein; and reform accordingly.
Not to talk much, nor of myself.
Not to boast of my former beauty, or strength, or favour with Ladies, etc.
Not to hearken to Flatteries, nor conceive I can be beloved by a young woman.
Not to be positive or opinionated.
Not to sett up for observing all these Rules, for fear I should observe none.

Further Reading

Banks, J., et al., eds. (2008), *Living in the 21st Century: Older People in England*. English Longitudinal Study of Ageing (Wave 3), Institute for Fiscal Studies.

de Beauvoir, S. (1970), *The Coming of Age*. Norton.

Carstensen, L. L., and Hartel, C.R., eds. (2006), *When I'm 64*. National Academies Press.

Cabeza R., et.al., eds. (2005), *Cognitive Neuroscience of Aging: Linking Cognitive and Cerebral Aging*. OUP.

de Grey, A., and Rae, M. (2007), *Ending Aging*. St Martins Press.

Jacoby, R., et.al. (2008), *The Oxford Textbook of Old Age Psychiatry*. OUP.

Johnson, M. L., et al (2005), *Handbook of Age and Ageing*. Cambridge University Press.

Kirkwood, T. (1999), *The Time of Our Lives: The Science of Human Aging*. OUP.

Kurtz, I. (2009), *About Time: Growing Old Disgracefully*. John Murray.

Magnus, G. (2009), *The Age of Aging: How Demographics are Changing the Global Economy and Our World*. John Wiley.

Marmot, M. (2005), *The Status Syndrome: How Social Standing Affects Our Health and Longevity*. Owl.

Morley, J. E., 'A Brief History of Geriatrics'. *Journal of Gerontology* 2004, 59A, 1132–52.

Neuberger, J. (2009), *Not Dead Yet: A Manifesto for Old Age*. Harper.

Partridge, L., Thornton, J. and Bates, G. (2011), *The New Science of Ageing*.

Philosophical Transactions of the Royal Society, 388, 6–7.

Sierra, F., et al. (2009), 'Prospects for Lifespan Extension'. *Annual Review of Medicine*, 60: 457–69.

Skinner, B. F., and Vaughan, M. E. (1983), *Enjoy Old Age*. Norton.

Silverstone, B., and Hyman, H. (2008), *You and Your Aging Parent*. OUP.

Index